CO-OPERATION AND DEVELOPMENT IN THE ENERGY SECTOR

Co-operation & Development in the Energy Sector

THE ARAB GULF STATES AND CANADA

Edited by Atif A. Kubursi and Thomas Naylor

Proceedings of a Symposium on the Energy Sector
co-sponsored by the Petroleum Information
Committee of the Arab Gulf States and McMaster
University, Canada, and held at McMaster University
16–17 May 1984

CROOM HELM
London ● Sydney ● Dover, New Hampshire

© 1985 Petroleum Information Committee of the Arab Gulf States
Croom Helm Ltd, Provident House, Burrell Row,
Beckenham, Kent BR3 1AT
Croom Helm Australia Pty Ltd, Suite 4, 6th Floor,
64–76 Kippax Street, Surry Hills, NSW 2010, Australia

British Library Cataloguing in Publication Data
Co-operation and development in the energy sector:
  the Arab Gulf States and Canada: proceedings of
  a symposium on the energy sector co-sponsored by
  the Petroleum Information Committee of the Arab
  Gulf States and McMaster University, Canada and
  held at McMaster University 16–17 May 1984.
  1. Petroleum industry and trade —— Political
  aspects   2. International economic relations
  I. Petroleum Information Committee of the Arab
  Gulf States   II. McMaster University   III. Kubursi, Atif
  IV. Naylor, Thomas
  338.2'7282   HD9560.5
  ISBN 0–7099–1582–9

Croom Helm, 51 Washington Street, Dover,
New Hampshire 03820, USA

Library of Congress Cataloging in Publication Data
Symposium on the Energy Sector (1984: McMaster
    University)
    Co-operation and development in the energy sector.

    Proceedings of a Symposium on the Energy Sector
co-sponsored by the Petroleum Information Committee of
The Arab Gulf States and McMaster University, Canada
and held at McMaster University 16-17 May 1984.
    1. Petroleum industry and trade — Persian Gulf
Region — Congresses.   2. Petroleum industry and trade —
Canada — Congresses.   I. Kubursi, A.A.   II. Naylor,
Thomas.   III. Petroleum Information Committee of The
Arab Gulf States.   IV. McMaster University.   V. Title.
HD9576.P52S95   1984   333.8'232'09536   85–12840
ISBN 0–7099–1582–9

Filmset by Mayhew Typesetting, Bristol, England
Printed and bound in Great Britain
by Mackays of Chatham Ltd

CONTENTS

ACKNOWLEDGEMENTS

This volume groups together the proceedings of a conference organised in Hamilton, Ontario from 16–17 May 1984 and co-sponsored by McMaster University and the Petroleum Information Committee of the Arab Gulf States.

The success of the symposium and the completion of this volume owe a great debt to His Excellency Sheikh Nasser Al-Ahmed Al-Jaber Al-Sabah, the Chairman of the PIC. The organising committee both at McMaster and in Kuwait contributed substantially to this success. Particular thanks are due to Dr Walid Sharif, Professors Byron G. Spencer, Mohamed A. Dokainish, Syed Ahmed, Mr Edmund Shaker and Mr Al-Feili. Thanks are also due to all those who participated in the symposium. Their comments and contributions added to the lively discussions that characterised the proceedings.

We are particularly grateful to Betty May Lamb for typing and organising this manuscript and for her dedicated contribution to the organisation of the symposium.

INTRODUCTION

Atif A. Kubursi and Thomas Naylor,
McMaster University and
McGill University, Canada

This is a collection of essays inspired by a mixture of concern and hope. The concern derives from the appalling — and unfortunately growing — disparities of income and wealth between countries and among their various social and regional subdivisions. The hope derives from the realisation that the first step towards rectifying at least some of the causes of concern lies in understanding their origins.

The central theme of this collection of essays is that, despite the glaring contrasts of wealth and poverty, expectation and despair, knowledge and ignorance, and above all else power and impotence, or perhaps because of them, there is a pressing need for a deeper understanding of certain fundamental economic structures — without which co-operation and collaboration in the construction of a more equitable framework for international economic relations will not be possible.

This volume is intended as a modest step in that direction. It consists of 7 papers presented at an international symposium on the energy industry held at McMaster University, in Hamilton, Ontario on 16–17 May 1984. The symposium was co-sponsored by McMaster University and the Petroleum Information Committee of the Arab Gulf States. Among the contributors to this volume are some of the most influential policy makers in the Gulf States and in Canada, along with some of the leading petroleum economists in these countries.

The principal contribution of this volume lies in the fact that it expresses the views of both oil-producers and oil-consumers, and details in a systematic way some of the energy-related problems they have encountered as well as some of the solutions they have devised. What emerged from this discourse is a clear picture of their communality of interest in the conservation and judicious management of the dwindling world supply of non-renewable energy sources.

The first part of this volume consists of the opening statements of Sheikh Nasser Al-Ahmed Al-Jaber Al-Sabah, and of Mr C. Geoffrey

3

Edge, Chairman of Canada's National Energy Board. Sheikh Nasser focussed attention on the dialogue between oil producers and oil consumers, and highlighted the interdependence between Canada and the Arab world, as well as the scope for widening the fields of co-operation between them. His main theme however is the urgent need to improve relations between North and South, in order to defuse the existing political tensions and to preserve international peace and security.

On the other hand Mr Edge stressed that we all need to learn from each other's experience in order to draw profit from the interdependence that we all face.

Next Dr Abbas Al-Nasrawi opens the main part of the proceedings with an overview of the current state of the world oil market in light of past trends and pressures. He stresses the political constraints within which OPEC must operate, with the result that the market, rather than cartel-type pressure, is the ultimate arbiter of oil prices. That in turn accounts for the recent weakening of oil prices, and consequently of the financial position of the OPEC countries.

The contribution of Dr L.C. Atkinson takes the survey of the state of the world oil market from the present into the future. He advances a number of reasons for supposing that oil prices will be likely to stabilise at their current values for a few years. Among them are declines in the rate of inventory drawdowns and expected increases in demand for oil as the major consuming countries emerge from the existing recession.

The issue of optimal oil pricing in an uncertain world is the central theme of the contribution of Dr Abdullah El-Kuwaiz. He points out two important lessons learned by the Gulf producers. One is that the availability of foreign financial assets has offered a modicum of stability to national income flows during times when world oil demand was weak. The second is that, at least in the short run, increases in oil demand lead to greater benefits to the producers when output is adjusted to meet those increases, rather than when prices alone are raised.

The social impact of higher oil revenues on the Gulf region is the subject of the contribution of Dr Mohammed Al-Rumaihi, presented at the luncheon banquet. He drew a comprehensive picture of the positive and the negative social consequences of the rapid pace of economic development; and he highlighted the importance of the Gulf Co-operation Council in providing an optimal framework for regional development.

Drawing on a wealth of practical experience within the Algerian and the broader OPEC contexts, Dr Maachou, the next contributor, stressed that energy policies are formulated — or should be formulated — within a broader framework given by a country's development strategy and the global economic (and political) context. Dr Maachou also stressed that the Gulf producers in particular were highly sensitive to world requirements in formulating their pricing and output decisions, more sensitive in fact than to their own immediate economic needs. In his opinion, they had received little back in return for this sacrifice; and he called for a new spirit of co-operation and dialogue that would be consistent with justice, equity and long term stability.

Considerable discussion of the contributions dealing with the world oil market and the Gulf States' strategy followed. The next part of the conference proceedings deals with the Canadian context.

Canada is in a sense a microcosm of the world of oil. The country is simultaneously an exporter and an importer, the western part of the country marketing its surpluses in the US, while in the east of Canada the market is served by imported oil. The result has been a complex of fiscal transfers from the federal government to the provincial governments of the oil-importing provinces, while the federal government has attempted to finance those transfers, at least in part, by tapping the revenue flows from the oil exported out of the producing provinces towards the US. Thus, in common with economists and planners in the Gulf States, those of Canada are concerned with equity and efficiency, national sovereignty and inter-regional distributive justice, the need for a 'fair' price of oil for both consumers and producers, and the economic consequences of the resource transfers that occur when real oil prices change substantially.

Therefore it is not surprising that the section of the conference dealing with Canadian energy policies and problems featured a debate between Dr Paul Tellier of the federal government, defending the government's National Energy Policy on the one hand, and, on the other, Dr Brian Scarfe of the University of Alberta (situated in Canada's principal oil-producing province) who subjected that National Energy Policy to searching criticism.

According to Dr Tellier, the NEP (as it is called in Canada) was designed against the background of the 1970s when insecurity of world supplies seemed to prevail, in order to assure that security of supply for Canada. But the NEP also attempted to assure for the various levels of government a better sharing in the revenues derived from rising oil prices. And it had as another objective, increasing Canadian equity participation in a Canadian industry traditionally dominated by the large international oil companies.

Dr Scarfe attacked both the objectives and the results of the NEP. In his view it has harmed the cash flow of the industry, reducing investment and exploration, alienated foreign capital, and rigidified the pricing structure. It may also have had a hand in perpetuating inefficiencies in Canadian energy-consuming industries by artificially holding down the price.

Considerable discussion followed this part of the conference proceedings, with questions focussing on the interaction between the Canadian energy situation and its overall foreign policy stance towards the Middle East.

The proceedings of the seminar concluded with a banquet speech by Sheikh Nasser, and kind words of appreciation and a documentation of the intellectual links between the West and the Arabs by McMaster's President Dr Lee.

OPENING ADDRESS

His Excellency Sheikh Nasser Mohammed
Al-Ahmed Al-Jaber Al-Sabah,
Minister of Information, Kuwait

Ladies and Gentlemen

It is indeed a great pleasure for my colleagues and myself, representing Saudi Arabia, Iraq, UAE, Qatar, Bahrain, Oman and Kuwait, to be with you today at this distinguished University.

I would like to express my sincere gratitude to McMaster University for its kindness in co-sponsoring with the Petroleum Information Committee of the Arab Gulf States, this symposium on the energy industry.

I would like also to take this opportunity to thank the organising committee of today's symposium.

I cordially welcome all participants, who are meeting today to explore common experiences and mutual interests.

Our meeting at McMaster University today represents a most welcome opportunity for conducting a fruitful dialogue between the developing oil-producing countries of the Arabian Gulf and Canada, a fully developed and industrialised oil-producing and oil-consuming country of the West.

We believe that in spite of the many differences between our distant regions, there exists among us a great potential for co-operation which has yet to be exploited fully. This potential stems from our complementary economies, from our traditional friendship and, not least, from our common interest in promoting greater prosperity, security and peace throughout our regions. Our desire to achieve wider co-operation therefore transcends purely commercial motivations and reflects our close interdependence and mutual concerns.

We firmly believe that there is ample room for a fruitful economic, technological and cultural exchange between us. This symposium is a modest gesture in this direction. We need much more.

Ladies and Gentlemen: today we are all too aware of the fact that our present world is divided into two large and drastically different

9

socio-economic regions. First, there are the industrialised countries possessing technology and bountiful resources, including oil. Second, there are the countries of the Third World, some of them, like us in the Arabian Gulf, fortunate to possess oil or some other raw material, but still lacking technology and other developmental prerequisites. There are also a large group of unfortunate countries that possess neither technology nor raw materials. The existing imbalance between these regions is generating tensions which could easily lead to very dangerous and undesirable consequences for the world community at large.

Until very recently, the idea of interdependence between these two worlds was not generally acceptable. Today, however, it is clear to all of us that an improvement in the relations between these two regions is required immediately to defuse the existing political tensions and to preserve international peace and security. Our interdependence in turn means that we have a collective responsibility for the efficient and equitable use of all our resources: human, technical and natural, including, of course, oil.

We are still far away from this rational order of things. The prevailing mechanisms governing the relationship between the oil-producing and oil-consuming countries are such that another energy crisis may not be avoidable.

In order to define the problems with which we are dealing, we must start to redefine the consumer–producer relationship. The oil-consuming countries have not acknowledged fully the new reality which has emerged since the early 1970s. This new reality has confronted both the consumers and the producers with a series of political and economic challenges which, because of imperfect knowledge and a narrow perspective on the issues, were not always fully met. Therefore, our objective must be, first and foremost, to offer a deeper knowledge of our respective economic realities, of our reciprocal interests, and of our common responsibilities for the stability and prosperity of the world community, in order to arrive at a clearer identification of our respective roles in reaching our common goals.

We, the oil-producing countries of the Arabian Gulf, see ourselves as members of the Third World. We therefore welcome international co-operation to assist us in implementing our difficult development programmes, as long as this assistance does not increase the economic costs to our development process, or infringe our political and economic independence, or deplete unduly our natural resources.

Although our past experiences with international co-operation, especially with the industrialised countries, have shown us that we were all too often subjected to unacceptable political and economic pressures, we are still committed to more co-operation. None the less we need to ensure that this co-operation does not reproduce the injustices of the past.

I would like to take this opportunity to single out Canada as a major leader in promoting world peace and understanding. We in the Arab Gulf States are indeed heartened and appreciative of your government's continued efforts for increased world dialogue and co-operation. We are privileged and delighted to co-operate with you in this effort; and we view our participation in this symposium here in Canada as a modest expression of our commitment to, and appreciation of, your efforts in this regard.

As you know, the Gulf war has been raging for the past four years with heavy material and human losses on both sides. Our brothers in Iraq have called repeatedly for the end of this savage war, to no avail. The other side refuses all peace initiatives. We hope that we can all join efforts to end this conflict peacefully.

Ladies and Gentlemen: I would like on behalf of all my colleagues, and the people of the Arab Gulf, to wish you a successful and productive symposium.

OPENING REMARKS

Mr C. Geoffrey Edge,
Chairman, National Energy Board,
Ottawa, Canada

Your Excellencies, Mr Chairman, I am deeply honoured to have been invited to deliver some opening remarks to this distinguished gathering for the Fifth Symposium of the Petroleum Information Committee of the Arab Gulf States.

I am told that the previous four conferences have been held in various locations in Europe. We are delighted that you chose Canada as the location for your first meeting on this side of the water; and I am sure I speak on behalf of all my fellow countrymen when I tell you how pleased we are to have the opportunity to participate on this occasion.

Since Paul Tellier, the Deputy Minister of Canada's Department of Energy, Mines and Resources, is scheduled to speak to you later about this country's national energy policy, I will try to steer clear of his territory by adhering to a more general frame of reference.

May I say at the outset that I applaud the aims of your Committee, which — as I understand it — include the promotion of a dialogue between the producing and consuming nations on all relevant matters involving petroleum and related energy sources.

Since Canada is a major producer of many different types of energy — oil, natural gas, elecricity and coal in particular — and since it is also one of the highest *per capita* consumers of energy in the world, we have considerable interest in these issues from both perspectives.

I must acknowledge that I found the title for this symposium — Canada and the Arab Gulf States: Different Pasts, Shared Prospects — to be both very intriguing and thought-provoking.

Our pasts have, indeed, been very different and, if truth be told, our respective knowledge about each other has, until quite recently, probably been minimal. While your history, culture and traditions stretch back into the mists of antiquity, most of us — the native people of Canada excepted — have roots in this country going back no more than a few centuries at most. While most of the inhabitants in earlier years were descended from French and British immigrants, later citizens arrived from many other countries in Europe and around the world.

Economically, Canada has long formed a part of the Western industrial world and evolved in line with its developments. You in the Arab Gulf States followed a different path, embarking only in recent years on an industrial revolution. I am reminded of a visit I made to the small walled city of Kuwait in 1942 where life went on at the same unhurried and uncomplicated pace as it probably had for a thousand years. I am sure I would not recognise the modern Kuwait of today.

On the basis of experience of the past decade, I have no doubt about the shared prospects of Canada and the Arab Gulf States. Certainly we in this country have been profoundly affected, as have most other nations, by the far-reaching changes that have been taking place over the past several years in relation to almost every aspect of energy, including such fundamentals as alternative sources of supply, demand and price.

I expect that Dr Tellier and other Canadian speakers will indicate in some detail the enormous impact that international developments of the past several years have had on almost every aspect of energy development in this country. I suppose it is also true to say — and perhaps we will have an opportunity to develop a better understanding on this point — that developments that have taken place in Western industrial countries such as Canada, including the problems we experienced of soaring inflation and severe recession, have also had a major impact on the oil-producing states of the Gulf area.

What I think has become very apparent from these developments to which I have just referred is the extent to which all of us are becoming increasingly interdependent in a world that, in every other way but geographically, is becoming smaller and smaller every day.

We are all affected by the actions of the superpowers. Environmental problems know no boundaries, as we are learning to our distress both in North America and in Europe. We cannot isolate ourselves from the problems of the Third World because ultimately they are also our problems.

While their relative importance varies from country to country, in most nations developments involving energy can have a far-reaching effect on the general course of their economies. Present and prospective future levels of world oil prices, for example, can have a profound effect on supply, demand and on investment decisions, which in turn can have a major impact on the world economy.

I believe that we have much to learn from the experience of other countries, to the mutual advantage of all. This point was impressed on me very strongly only this past Monday, when I had the opportunity to participate in the opening of Canada's second international energy conference, known as Energex '84, at Regina, Saskatchewan. This week-long gathering brought together several hundred experts in the field of energy from several countries around the world.

That these two gatherings should be taking place at almost the same time in two of our cities, perhaps attests better than anything to the somewhat belated lesson we have learned, namely that we are dependent one upon the other. If we can keep that lesson in mind, I think there are opportunities to be seized to our mutual benefit.

I

OPEC AND THE CHANGING STRUCTURE OF THE WORLD OIL MARKET

Dr Abbas Al–Nasrawi,
Professor of Economics,
Department of Economics,
University of Vermont, USA

The Organization of the Petroleum Exporting Countries (OPEC) was created in 1960. It was conceived by its founding members (Saudi Arabia, Venezuela, Iraq, Iran and Kuwait) as an instrument to pressure the multinational oil companies to raise the member governments' revenue, which at the time was about 85 cents per barrel. When, in 1973, OPEC succeeded in sharply raising the price of crude oil and, therefore, the revenue of its member countries, it began to be perceived as a monolithic oil power capable of unilaterally setting oil prices and determining output levels. This perception was reinforced by the successful effort by the Arab members of OPEC to reduce oil output in the context of the October 1973 Arab-Israeli War. Such a perception is not supported, however, by the reality of OPEC behaviour since its inception.

For one thing, prior to 1973, OPEC had failed to keep the price of crude oil in line with the prices of other primary products and raw materials. More importantly, however, is the fact that OPEC adheres to the premise that member countries may do as they wish. Since they are sovereign states, they are not obligated to submit to the dictates of an intergovernmental organisation such as OPEC. This means that, in the last analysis, it is the market, and not OPEC, that will determine how much oil each member country will be able to sell.

In addition to acceptance of the concept of the sovereignty of the state, there is another factor which tended to reduce the power of OPEC in the international oil industry. Prior to 1973, control over oil output and price decisions in OPEC member countries were exercised by a very small number of large multinational oil corporations. These companies developed an elaborate system of output control mechanisms to funnel oil from different sources of supply into their world-wide network of integrated operations. The transfer of controlling power in the post-1973 era was *not* accompanied by OPEC adopting an output control mechanism to replace the one which had so effectively served the oil companies.

21

The failure to relate output from member countries to the world oil demand was not important so long as the demand for OPEC oil was high enough to generate the necessary financial resources needed by member countries. But as soon as there was a downward movement in demand for oil, OPEC's weakness as a price administrator was revealed. Such weakness was unavoidable since OPEC, like any other seller, could not control both the price charged and the quantity bought. This was particularly true in the case of OPEC since member countries retained for themselves the exclusive right to regulate the volume of output. This fragmentation of control over output decisions meant that the cartel function, which the multinational oil companies exercised prior to 1973, ceased to be effective in the post-1973 era.

The replacement of company control by individual state decision meant also a change in the very purpose of control. Under the company system, the purpose can be said to have been profit maximisation. This cannot be said about the state control system since oil became an integral part of the overall economic, social and political goals of the state. And whatever power OPEC was capable of exerting *vis-à-vis* the oil companies was dissipated among the OPEC member countries. The lack of control over the oil industry by OPEC was driven home at the time the second oil price shock of 1979–81 made its impact felt on the demand for oil. The higher price of oil, and the steep and long recession in the industrial countries, made it clear that the demand projections which had been made only a few years earlier had been rendered worthless by the realities of market conditions. Having failed to replace the output control mechanism of the oil companies by a supply regulating mechanism of their own, it was inevitable that certain market conditions would emerge that would be beyond the control of OPEC. These factors include certain serious structural changes in the international energy economy, certain structural changes in the international oil industry, and the drive of the industrial countries to lessen their dependence on OPEC member countries as the major sources of imported oil. These changes and their implications for OPEC will be examined in the following pages.

Structural Changes in the Energy Market

The significance of oil and its rising importance for the industrial economies may be appreciated by examining some economic indicators. Taking the member countries of the Organization for Economic

Table 1.1: Total Energy Consumption and Total Oil Consumption by Country Groupings, 1950–1982

	1950	1960	1970	1975	1980	1982
Industrial Countries						
Total Energy Consumption (MBDOE)	26.7	37.7	65.5	70.2	76.3	72.0
Oil Consumption (MBD)	8.3	15.5	33.2	36.9	38.0	34.1
Share of Oil in Total Consumption (%)	31.0	41.0	50.7	52.7	49.8	47.4
Centrally Planned Economies						
Total Energy Consumption (MBDOE)	7.8	18.8	27.5	35.5	42.8	44.6
Oil Consumption (MBD)	1.2	2.9	7.0	10.5	12.7	12.8
Share of Oil in Total Consumption (%)	15.4	15.4	25.4	29.6	29.7	28.7
Developing Countries						
Total Energy Consumption (MBDOE)	3.1	6.0	11.0	14.2	18.2	20.8
Oil Consumption (MBD)	1.2	3.2	6.3	8.3	11.0	11.6
Share of Oil in Total Consumption (%)	38.7	53.3	57.3	58.5	60.4	55.8
World Total						
Total Energy Consumption (MBDOE)	37.5	62.4	103.9	119.9	138.3	137.4
Oil Consumption (MBD)	10.8	21.6	46.4	55.7	61.7	58.5
Share of Oil in Total Consumption (%)	28.8	34.6	44.7	46.5	44.6	42.6

Source: OECD, *Energy Policy Problems and Objectives* (Paris, 1966); British Petroleum, *BP Statistical Review of the World Oil Industry* (annual); OPEC, *Annual Statistical Bulletin*.

Cooperation and Development (OECD) as a group we find that their total energy consumption increased from 27 million barrels per day of oil equivalent (MBDOE) in 1950 to 66 MBDOE in 1970, an increase of 162 per cent (see Table 1.1). Within this total, oil consumption increased by over 300 per cent, or from 8 MBD to 33 MBD during the same period. To put the rising importance of oil in perspective, it can be observed that, in 1950, oil supplied 31 per cent of total energy consumption in the OECD countries. Ten years later oil increased its share to 51 per cent. By 1973 it reached its peak when it supplied 56 per cent of total energy consumption. During the same period, 1950–73, the combined gross domestic product (GDP) of these countries increased from $1.4 trillion to $4.1 trillion, an increase of 193 per cent.

The relationship between oil consumption and GDP, or the energy intensity ratio, provides another indicator of the rising importance of oil. In 1950 energy requirements per $1,000 of real GDP in the OECD countries amounted to 6.6 barrels of oil equivalent. By 1973 this ratio had declined to less than 6 barrels of oil equivalent. By contrast the oil intensity ratio for the same group of countries has increased from 2.13 barrels of oil equivalent per $1,000 of GDP in 1950 to 3.35 barrels in 1973.[1] It is clear from the data that the ratio of growth of oil consumption considerably exceeded the rate of growth of GDP. This disparity in the growth rates reflected the fact that the increase in energy consumption was met primarily by increases in the consumption of oil. Another explanation for the increase in the share of oil in the energy market was the interfuel substitution in favour of oil due to environmental considerations in the industrial countries in the 1960s and the 1970s.

It is important to note that the trend towards the rising share of oil in the energy market was helped considerably by oil's own low cost relative to other forms of energy. This was particularly the case during the decades prior to 1970 when the revenue of the governments of the oil-producing countries was about 85 to 90 cents per barrel. And if we were to take into account the constant erosion in the terms of trade of the OPEC member countries during that period, it can be concluded that the real cost of oil to the OECD economies was negligible.

But the quadrupling of the price of crude oil in the early 1970s and the transfer of controlling power to the oil-producing countries led eventually to a restructuring of the energy market. The sharp rise in the nominal cost of oil to the OECD countries made the cost of oil as a variable input an important cost factor. This, together with the resultant balance of payments difficulties and concerns about security of supply, led the major oil-consuming countries to adopt new energy policies that would reduce the consumption of energy in general, and lessen their dependence on imported oil in particular. The position of the OECD countries was articulated in a policy statement by the International Energy Agency as follows:

> The two oil crises have illustrated the threat which energy can represent to the economy. But they have also awakened the world to the challenge of structural adjustment in the energy economy. Structural change is necessary to remove energy as a major constraint to sustain non-inflationary economic growth and to reduce oil supply vulnerability. The main objectives of structural change are to use energy more efficiently and to reduce dependence on oil so as to achieve a more balanced mix of energy sources that is less exposed to supply insecurity.[2]

To achieve their objectives OECD countries adopted and implemented several energy policies including energy conservation; promotion of the development of alternative sources of energy; an increase coal utilisation strategy; adoption of minimum energy efficiency standards; promotion of nuclear generating capacity; and the progressive reduction of oil as a source of electric power.[3]

In assessing the effectiveness of these policies in the OECD countries during the period 1973 to 1980, the IEA concluded that the following changes in the energy market have taken place.[4]

(1) Real GDP increased by 19 per cent but the increase in total energy requirements was only 4 per cent reflecting a significant historical reversal in the relationship between energy consumption and GDP growth. This historical change was reflected in the 13 per cent decline in the energy used to produce one unit of GDP.

(2) More significantly from OPEC's perspective, the successful energy policies of the OECD countries resulted in a reduction in their oil imports from all sources by 14 per cent and the oil used to produce one unit of GDP declined by 20 per cent.

(3) Domestic energy production increased by 13 per cent with oil rising by 9 per cent, coal by 23 per cent and nuclear energy by 206 per cent.

These changes led to a drastic reversal in the role of oil in the energy market. From being responsible for 56 per cent of total energy consumption in 1973, the share of oil declined to 45 per cent by 1982, again reflecting a shift in favour of non-oil sources of energy. It should be noted also that the efficient utilisation of energy was reflected in a decline in energy requirements in all segments of the energy market — industrial production, space heating and transportation.[5]

OPEC and the Structural Changes in the Oil Market

In addition to the structural changes in the energy market, and therefore in the oil market, there were also changes within the international oil industry which led to a drastic change in the position of OPEC member countries within that industry.

One such change was the increase in the number of sellers in the oil market. This was related directly to the expanded role of governments in the oil sector after 1973. This rise in the number of national oil companies, created after 1973 to replace the multinational oil corporations in marketing, increased markedly the supply of oil.

The sharp rise in the price of crude oil, and the consequent balance of payments difficulties, led many oil-importing countries to secure favourable terms from oil sellers. The increase in the number of sellers and the number of oil buyers had the effect of increasing competition in the oil industry.

Another change in the international oil market was the increase in the number of oil- producing and exporting countries outside the OPEC region. This was made possible by the sharp rise in the price of crude oil, which made it feasible to boost output from existing high cost oil areas. Thus the expansion in the output of oil in the United States, Mexico, Egypt, the United Kingdom and Norway had the effect of reducing the demand for OPEC oil. This may be seen in the decline in the share of OPEC oil in the world oil supply from 54 per cent in 1973 to 32 per cent in 1983. This change in the relative position of OPEC oil can be seen in the decline of OPEC oil output from 31.5 MBD in 1979 to 18.3 MBD in 1983. By contrast the oil output outside the OPEC region has increased during the same period from 34.3 MBD to 38.1 MBD at a time when the combined output of the OPEC and the non-OPEC region has declined from 63.8 MBD to 56.4 MBD (see Table 1.2).

OPEC and the 1983 Price Reduction

Since the 1973 price increase there was a general belief that an oil

Table 1.2: OPEC: Crude Oil Output 1961–1983 (million barrels per day)

Year	OPEC	World Total	OPEC Share in World Total (%)
1961	9.6	22.3	42
1970	23.4	45.7	49
1974	30.7	56.1	52
1979	31.5	65.8	48
1980	27.5	62.8	44
1981	23.3	59.2	39
1982	19.9	57.0	35
1983	18.3	56.4	32

Source: British Petroleum, *BP Statistical Review of the World Oil Industry* (annual); *BP Statistical Review of World Energy* (annual); OPEC, *Annual Statistical Bulletin*.

shortage was in the making and that the 1980s would witness a serious imbalance between oil supply and demand that would favour the oil-producing countries. This belief led to another, that the world community should curb its demand for, and dependence on, oil. The main concern in the 1970s was not whether prices would increase — that was taken for granted — but whether OPEC member countries would be able to supply the world with enough oil to meet its needs.

The belief that an oil shortage would emerge in the 1980s was based on a set of projections which assumed relatively high rates of economic growth in industrial as well as in developing countries. The fear of an oil shortage was strengthened by the drastic reduction in Iran's oil output and export due to that country's revolution. Indeed the belief that an oil shortage was unavoidable was so prevalent, that OPEC's own Long Term Strategy Committee concentrated its policy recommendation on how to adjust prices upwards.

OPEC's preoccupation with prices, it should be noted, was not a new phenomenon. The creation of OPEC was itself an attempt to increase revenue from oil through price increases. The 1973 price rise, which resulted in massive increases in member country financial resources. forced those countries to concentrate on how to spend and invest the financial resources — rather than on how to regulate output. This concern was reinforced by the fact that most energy studies were projecting a continued rise in the demand for oil.

It can be said that OPEC member countries would have felt the effect of the decline in the demand for energy and for oil before the end of the last decade, were it not for the political upheaval which engulfed the Middle East. In the last quarter of 1978 and the first quarter of 1979 the Iranian Revolution caused a severe decline in Iran's oil output and export. This decline prompted other oil-exporting countries in the Middle East to increase their output to offset it. But these acts had the effect of masking the developing problem of excess oil supply.

So too did the decision of the OECD countries in 1979 and 1980 to increase their oil inventories, a decision which had the effect of increasing these countries' demand for OPEC oil. This in turn led to a continued rise in the spot market prices which in time were followed by OPEC official prices. The panic buying, which had been triggered by the events of the Iranian Revolution, and which caused the rise in oil prices, continued until 1981.

The sharp rise in OECD oil inventories proved to be both costly and unnecessary, in the face of the world-wide decline in economic activity and unusually high interest rates. Furthermore, OECD energy conservation measures were succeeding in reducing the demand for energy, and for OPEC oil.

These conditions made it clear to OPEC countries in 1981 that further increases in the price of crude oil could not be justified by the prevailing market conditions.

The realisation that OPEC could not raise prices, coupled with the high cost of holding excessive oil inventories, led to a process of reintroducing these inventories into consumption channels. In other words in 1981 these inventories began to constitute another source of oil supply competitive with OPEC oil. This displacement of OPEC oil was accelerated by the decline in the demand for energy; by the substitution of other forms of energy for oil; and by the rise in the supply of non-OPEC oil. The combined effect of these changes was estimated to have reduced the output of OPEC oil from 31 MBD in 1979 to 19 MBD by 1982. The distribution of this reduction was estimated to be as follows:[6]

Decline in demand for energy	3.5 MBD
Displacement by non-OPEC oil	2.0 MBD
Substitution by non-oil fuels	3.0 MBD
Reduction in inventories	3.0 MBD

In order to protect their shares of the market, some OPEC member countries found themselves resorting to hidden or open price discounting. But such discounts proved to be insufficient to stabilise the oil price structure when the newcomers had no incentive to adhere to OPEC official prices. In order to protect and/or expand their share in a shrinking market, the North Sea producers resorted to a series of price reductions in 1982 and 1983, which brought the price of the North Sea oil from $36.50 a barrel to $30.50 a barrel. The $6 a barrel reduction brought the price of North Sea oil from its traditional level of being above the price of Arabian Light to a new level where it was $3.50 below the price of Arabian Light.

The reductions in the North Sea oil prices were important in their implications for OPEC. They confirmed that the market conditions had changed sufficiently that significant price reductions were necessary. They also reversed the traditional price structure where the prices of North Sea oil were maintained above the OPEC official price. Furthermore, the North Sea price reductions were followed by other price reductions by other exporters, such as Egypt, the Soviet Union and Mexico. Thus, the price setting power in the world oil market had moved to producers outside the OPEC region.

Following the North Sea price cuts, individual OPEC member countries such as Nigeria, Iran, Venezuela and Indonesia disregarded the collective price decisions by reducing their prices in an effort to protect their market shares. Other countries, which had continued to adhere to the official OPEC price structure, found themselves losing their market shares to the lower price producers, both outside the OPEC region and within. By early 1983 those member countries that were charging the official price realised that the situation was unacceptable. This in turn led in March 1983 to the first collective price cut in the history of OPEC.

The March 1983 decision by OPEC, which reduced the official price from $34 to $29 a barrel, also committed OPEC member countries to a combined output of 17.5 MBD in order to stabilise the new price structure.[7]

Concluding Observation

The 1973 price revolution, and the consequent transfer of control over oil from the oil companies to the oil-producing countries, created certain structural changes in the international oil industry.

The 1973 oil price revolution occurred at a time when demand for oil had been increasing for more than two decades. This led many observers to believe that there would be a shortage of oil some time in the 1980s. Neither forecasters nor OPEC policy-makers took very seriously the price elasticity of demand for oil. Energy policies adopted by the OECD countries led eventually to a reduction in energy and oil requirements. The second oil price shock of 1979–81 accelerated the tendency to reduce dependence on oil. The emergence of excess supply conditions prompted the non-OPEC oil-producing countries to reduce the price of their oil first. This was followed by price reductions by individual members of OPEC. In order to stem the destabilising effects of competitive price cuts, in March 1983 OPEC decided for the first time in its history to resort to a price reduction, and to agree to place ceilings on individual country output.

Notes

1. 'Energy Indicators', *OPEC Review*, vol. 6, no. 4 (Winter 1982), pp. 374–407.

2. International Energy Agency, *World Energy Outlook* (*WEO*) (Paris, 1982), p. 21.

3. The articulation of IEA's energy policies was expressed in a number of policy statements. For some of the more important pronouncements see *WEO*, pp. 46–59. See also IEA, *Energy Policies and Programs of IEA Countries* (Paris, 1977).

4. *WEO*, p. 22.

5. Ibid., pp. 91–9.

6. See The Research Group on Petroleum Exporters' Policies, 'Oil Prices in 1982: A Critical Year', *Middle East Economic Survey* (*MEES*), 6 December 1982 (Supplement), p. 7.

7. *MEES*, 21 March 1983, pp. A1–A8.

II

THE WORLD OIL MARKET FOR THE BALANCE OF THE 1980s: STABILITY OR TURMOIL?

Dr Lloyd C. Atkinson,
Senior Vice-President and Chief Economist,
Bank of Montreal

Recent Oil Market Developments

Spot crude prices were surprisingly firm well into the second quarter of 1984, as uncertainty about the Gulf War, declining Soviet exports, the prolonged coal strike in Britain, and technical difficulties in certain North Sea fields all exerted upward pressure on prices, thus helping to maintain the delicate equilibrium which has characterised world oil markets since the downward adjustment of the OPEC pricing structure in March 1983. More recently, however, downward pressures have intensified, confounding many oil market observers who had expected prices to escalate in the face of increased military activity in the Arab Gulf. Prices did not rise in May, at the escalation of hostilities. Since the end of May, however, prices have weakened, as the impact of fundamental supply and demand factors have overshadowed lingering fears about the Gulf War and possible supply shortages.

Prices for the Saudi Light benchmark crude have fallen nearly $1/barrel since the beginning of June, leaving prices almost $1.50 below official levels. Although prices for heavier OPEC crudes have also fallen since early June, they are still trading at a premium over official prices. Spot prices for higher quality African and North Sea crudes have varied substantially over the last 15 months, both above and below official prices, mostly in response to supply fluctuations. Over the last month, however, spot prices have declined rather steadily.

The recent slide in spot prices can be traced directly to OPEC oversupply. Although the drop in second quarter non-socialist world consumption (to about 44 MBD from an average 46.7 MBD in the first quarter) was largely expected, OPEC production in the second quarter is being maintained at roughly the first quarter average output rate of 18.1 MBD (excluding natural gas liquids). The primary reason for the continued strength of OPEC production can be traced to Saudi Arabia, which increased output levels from 4.6 MBD in March to 4.9 MBD in April and an estimated 5.4 MBD by June, to rebuild floating storage stocks and generally dispel market fears about possible supply shortages owing to escalation of the Iran-Iraq war in mid-May. By contrast, many other OPEC producers, including Nigeria, Algeria, Iran and Iraq have been exercising surprising restraint, producing at or close to their quota ceilings since April. While there are indications that production may have slipped to under 18 MBD by late June, largely as a result of reduced Saudi liftings, this is still well above the group's self-imposed 17.5 MBD ceiling, and well above the level needed to support a firming of prices.

None the less, it is our view that the current declining trend in spot oil prices is a temporary phenomenon which will be redressed, first, by cuts in OPEC production over the next few months and, ultimately, by a strengthening of demand in the final two quarters of the year. OPEC behaviour appears to be consistent with maintenance of the current pricing structure. Just as production stayed high in the second quarter to offset possible war-related shortages, it appears to be subsiding in the face of price weakness and reduced hostilities. Although it is still too early to tell, the most recent price data for early July suggest that the downward price slide may be losing some momentum.

Oil Market Developments Since March 1983

1. World Oil Prices

Underlying stability in spot market boosts confidence in $29/barrel pricing structure . . .

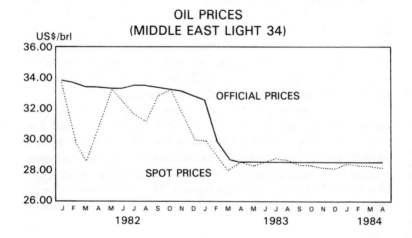

OIL PRICES (MIDDLE EAST LIGHT 34)

(a) Since the $5/barrel downward adjustment in OPEC's official pricing structure in March 1983, spot markets for most categories of OPEC crude have been remarkably stable. Although 'Middle East Light' (a weighted average of OPEC's light crudes) has been trading at a discount since last September, the differential from official prices has remained relatively narrow.

(b) After weakening in the 4th Quarter of 1983, prices firmed in early 1984, largely as a result of cold weather and an increase in demand for home heating fuels. Prices have fallen somewhat over the last few weeks reflecting seasonal factors associated with the normal 2nd Quarter downturn in demand.

(c) In early June, spot prices for Middle East Light appear to be fluctuating in the $28.20 to $28.40 range — as much as 80 cents below official levels. While some further seasonal weakening is possible, downward pressures are expected to ease by the 3rd Quarter as demand picks up and stocks are replenished in anticipation of winter demand.

(d) Prices for heavier OPEC crudes have remained persistently above official levels, reflecting economic recovery in the industrial world (particularly the US) and increased demand for heavy fuel oil. Reflecting the strengthening market for heavy crudes, Mexico recently (1 May) implemented a 50c/barrel increase on its own heavy (Maya) crude.

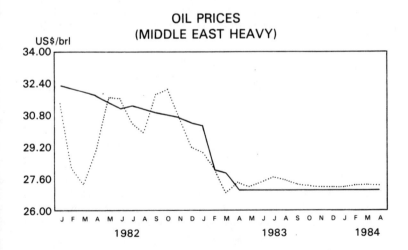

OIL PRICES
(MIDDLE EAST HEAVY)

(e) Since the weakness in the 4th Quarter of 1983, when spot prices for higher grade African and North Sea output were $1 or more below official levels, prices for these crudes have firmed significantly. This is attributable to economic recovery, the onset of colder weather in Europe in early 1984, and recent restraint in Nigeria's production, which earlier in the year had been running some 200,000 BD above its quote allotment.

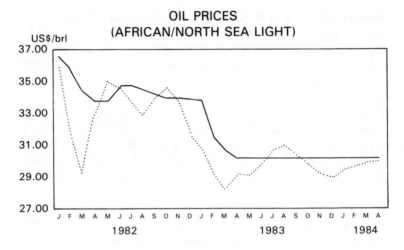

OIL PRICES (AFRICAN/NORTH SEA LIGHT)

(f) Another factor behind the firming of North Sea markets has been BNOC's adoption of a more flexible marketing strategy, involving increased activity in spot trading. This has strengthened the company's ability to withstand downward market pressures against term contract prices.

2. Demand and Supply

The stability in world oil prices is partly the result of a pickup in demand . . .

(a) Preliminary statistics for the 1st Quarter 1984 show non-socialist world demand for crude oil running about 700,000 BD ahead of 1st Quarter 1983 consumption. Cold weather and an associated surge in US demand have been largely responsible for the increase. For the year as a whole, this 700,000 BD differential is expected to be maintained, as a slowdown in the rate of US demand growth through the rest of the year is largely offset by a strengthening trend elsewhere in the industrial world. This suggests an average 1984 demand of 45.3 MBD, about 1.6 per cent above the 1983 average.

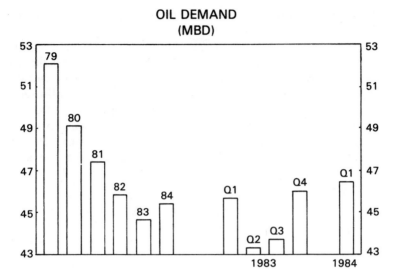

OIL DEMAND
(MBD)

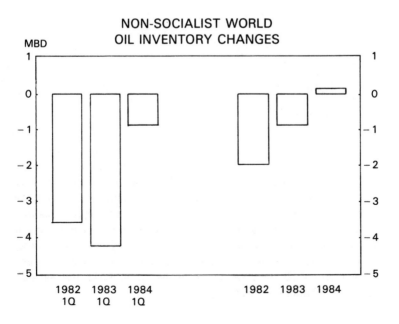

NON-SOCIALIST WORLD
OIL INVENTORY CHANGES

It is also partly the result of a slowing in the pace of inventory drawdown.

(b) The rapid pace of inventory reduction, which has characterised oil markets since 1982, appears to be subsiding. Supply and demand figures for 1st Quarter 1984 for the non-socialist world imply a stock reduction of just under 1 MBD. This is significantly less than the 3.6 MBD and 4.3 MBD drawdowns which occurred in the 1st Quarters of 1982 and 1983 respectively. For 1984 as a whole, inventory levels are expected to show little change from end 1983 levels, with drawdowns in the 1st and 4th Quarters offset by stockbuilding in late summer and early fall.

None the less, stocks remain high . . .

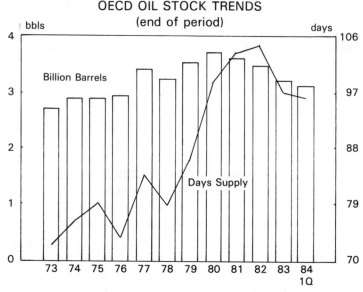

(c) Despite the significant reduction in OECD inventories which has occurred since the end of 1980 (nearly 650 million barrels), oil stocks remain high by historical standards. This reflects both the heavy accumulation of inventory in 1979 and 1980, and declines in consumption which have extended the stock in terms of days of supply. Important shifts have, however, taken place with regard to the allocation of inventories. Commercial stocks are now at their lowest level since 1975, when the IEA began collecting such data. Meanwhile, stocks held by governments (particularly the USA, Germany and Japan) have risen steadily, reflecting additions to strategic stockpiles.

Cumulative effect: rise in OPEC output compared to a year ago

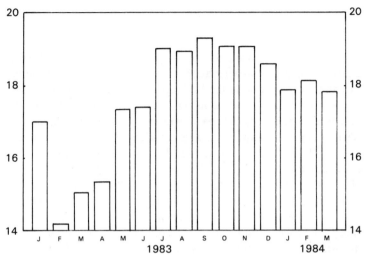

OPEC OIL PRODUCTION
(MBD)

(d) Increases in demand, and more importantly, the slowing of inventory drawdown, have resulted in strong output gains for OPEC producers. OPEC production averaged almost 18 MBD in the 1st Quarter 1984, nearly 2½ MBD above 1st Quarter 1983 output. Currently, production is running at close to 17.5 MBD. This is only slightly below 1st Quarter levels and is, in our view, consistent with continued price stability. The reason: the UK coal strike has boosted demand by some 500,000 BD, temporarily offsetting some of the seasonal decline normally expected in 2nd Quarter demand.

(e) For the year as a whole, OPEC crude oil production is expected to average just over 18 MBD, or 1 MBD greater than the 1983 average. Countries with a pressing need for higher oil revenues (e.g. Nigeria) have recently softened their demands for higher individual quotas, and formal changes in either individual quotas or the group's official 17.5 MBD output ceiling are not expected before the end of the year, at the earliest.

(f) Barring major escalation of the Middle East war, the current world supply-demand relationship for oil is such that no changes in the nominal

weighted official OPEC price of $28.60 are expected through the end
of 1984. This suggests a real price decline in the order of 5 per cent
for the year as a whole.

Longer-Term Outlook

Non-socialist world demand for oil is currently running about 1 MBD
ahead of 1983, as cold weather and economic recovery strongly boosted
1st Quarter consumption in the USA and Japan. Although the pace of
increase in demand is expected to slow through the remainder of 1984
in response to the expected slackening in economic growth, for the year
as a whole demand is expected to average 45.5 MBD, some 1.8 per cent
above the 1983 average. This will be the first annual increase in non-
socialist demand since 1979.

The current weakness in spot oil prices is expected to persist well
into the 3rd Quarter, because of the significant buildup in inventories
which has been taking place as a result of current overproduction.
Although the pace of 1st Quarter inventory depletion is now estimated
at 1.4 MBD, or 400,000 BD greater than originally projected, the 2nd
Quarter buildup appears to be averaging between 1 and 1.5 MBD. This
is about three times the normal 2nd Quarter rate. The bulk of this inventory accumulation appears to be finding its way into commercial stocks
in the USA and Japan. This should dampen demand for new production
in the 3rd Quarter when the bulk of pre-winter stockpiling usually occurs.

Between 1984 and 1988, demand for oil is expected to continue to
increase at an average annual rate of about 1.8 per cent. This is just over
one half the pace of increase projected for industrial world growth over
the same period. Such a pace of increase indicates non-socialist demand
of just under 49 MBD in 1988 which would just about restore demand
in that year to the 1980 level.

Reflecting the near term weakness in world oil consumption, demand
for OPEC oil is expected to weaken over the next few months, rising,
however, to over 19 MBD by the 4th Quarter. Because of the limited
possibility for achieving sustained increases in output from non-OPEC
sources, most of the medium-term increase in demand will be met from
OPEC production. Depending on expectations regarding imports and
natural gas liquids, our demand forecast suggests a demand for OPEC
oil betwen 22 and 23 MBD by 1988.

Barring major new developments in the Middle East war, the short-
term price weakness is not expected to precipitate a major downward

adjustment in the official OPEC price structure. No changes in the nominal weighted official average OPEC price of $28.60 are expected through the end of 1985. This implies continuing real declines in prices over the next two years. Between 1985 and 1988 nominal increases in oil prices are expected just to keep pace with industrial world inflation, implying no real increase and a nominal average weighted OPEC price of $32.60 in 1988.

III

THE FUTURE OF OIL PRICES

Dr Abdullah El-Kuwaiz,
Assistant Secretary General
for Economic Affairs,
The Co-operation Council of
the Arab States of the Gulf,
Deputy Minister of Finance,
Kingdom of Saudi Arabia

Recent history has taught all of us what we should have expected. Oil prices can go down as well as up. It has proven what we in OPEC have been saying all along, that it is market forces, as opposed to alleged collusive behaviour, that have shaped oil prices. For us in the Gulf, two important lessons can be drawn.

First, the availability of foreign assets offers flexibility when there is slack in the oil market.

Second, when the demand for OPEC oil increases, oil-exporting countries, especially those with excess installed capacity, can benefit more by exporting increased quantities to support prevailing prices, than they can from maintaining current low export levels at higher prices. Let me now discuss in more detail what these lessons from the past teach us about the future.

A glance at Table 3.1 will show that OPEC crude oil production hit a peak in 1977 at over 31 MBD; in 1982 it was 18.7 MBD; and estimates for 1983 are at about 18 MBD. Table 3.2 shows the maximum sustainable capacity figures for OPEC; for 1985, the aggregate figure is projected in the range of 31.6 to 33 MBD; and in 1990, between 29.1 and 34.3. This table indicates that OPEC countries could, if they wanted to, increase output by 16 MBD, with only maintenance investment.

Between now and 1990, the question as to what will happen to the oil price clearly depends on the demand for oil, which in turn depends on world economic growth; the income elasticity of the demand for oil; the price elasticity of the demand for oil; on the supply of oil and alternative energy sources in OPEC and in non-OPEC countries; and on OPEC's supply of oil. We can do a 'back of the envelope' calculation to arrive at some projections.

Projected world economic growth from the IMF and other sources would indicate optimistically an annual growth rate of less than 4 per cent. Estimates for long-run average income elasticity of demand for oil

45

Table 3.1: Crude Oil Production, 1973–1982 (thousand barrels per day)

	1973	1974	1975	1976	1977	1978	1979	1980	1981	1982
Iran	5,861	6,022	5,350	5,883	5,663	5,242	3,168	1,662	1,380	2,214
Iraq	2,018	1,971	2,262	2,415	2,348	2,563	3,477	2,514	1,000	972
Qatar	570	518	438	497	445	487	508	472	405	329
Saudi Arabia	7,596	8,480	7,075	8,577	9,245	8,301	9,532	9,900	9,815	6,470
United Arab Emirates	1,533	1,679	1,664	1,936	1,999	1,831	1,831	1,709	1,474	1,214
Kuwait	3,020	2,546	2,084	2,145	1,969	2,131	2,500	1,656	1,125	827
Algeria	1,097	1,009	983	1,075	1,152	1,161	1,154	1,012	805	710
Libya	2,175	1,521	1,480	1,933	2,063	1,983	2,092	1,787	1,140	1,158
Gabon	150	202	223	223	222	209	203	175	151	155
Nigeria	2,054	2,255	1,783	2,067	2,085	1,897	2,302	2,055	1,433	1,295
Indonesia	1,339	1,375	1,307	1,504	1,686	1,635	1,591	1,577	1,605	1,339
Ecuador	209	177	161	188	183	202	214	204	211	210
Venezuela	3,366	2,976	2,346	2,294	2,238	2,165	2,356	2,168	2,102	1,891
	30,988	30,731	27,156	30,737	31,298	29,807	30,928	26,891	22,646	18,784
World Total	55,674	55,852	52,880	57,312	59,685	60,057	62,535	59,538	55,900	53,162

Source: US Department of Energy.

Table 3.2: OPEC Crude Oil Productive Capacity (million barrels per day)

	1981 Actual Crude Oil Output	Maximum Sustainable Capacity		
		1985	1990	2000
Algeria	0.8	0.9	0.6–0.7	0.3–0.4
Ecuador	0.2	0.1	0.1–0.2	0.1–0.2
Gabon	0.2	0.1	0.1–0.2	0.0–0.1
Indonesia	1.6	1.3	0.9–1.4	0.8–1.2
Iran	1.3	4.0	3.6–4.0	3.0–4.0
Iraq	0.9	3.5	4.0–4.5	4.0–4.5
Kuwait	1.1	2.5	2.4–2.6	2.4–2.6
Libya	1.1	2.0	1.5–1.8	1.5–1.8
Neutral Zone[1]		0.6	0.4–0.5	0.4–0.5
Nigeria	1.4	2.2	1.6–2.2	1.2–1.63
Qatar	0.4	0.5	0.2–0.4	0.1–0.2
Saudi Arabia	9.8	9.5–10.5	9.5–11.0	9.5–11.0
United Arab Emirates	1.5	2.0–2.4	2.0–2.4	2.0–2.4
Venezuela[2]	2.1	2.4	2.2–2.4	2.4–3.0
Total OPEC	22.5	31.6–33.0	29.1–34.3	27.8–33.5

Notes:

1. 1981 output included in Saudi Arabia's and Kuwait's output.
2. Venezuela's productive capacity assumes 0.1–0.2 MBD of liquids from the oil belt in 1990 and up to 1.0 MBD in the high case for the year 2000.
Source: *World Energy Outlook*, International Energy Agency, 1983.

are in the range of 1.3 to 1.9 for LDCs and 0.8 for industrial countries. The long-run price elasticity for the LDCs is in the range of –0.25 to –0.35, and for the industrial countries it is in the order of –0.4.

The other element to be considered is oil production. As an extreme case scenario, we can assume a zero increase in production outside of OPEC. OPEC has roughly an additional 16 MBD that it can produce. Of this, it is estimated that an additional 2 MBD would go to domestic consumption by 1990, leaving 14 to 16 MBD (16 in 1985, going down to 14 in 1990) for exports; this is roughly sufficient to meet projected world demand if the world economy grows at the projected 4 per cent annual rate.

The question, therefore, becomes — how much will OPEC, in fact, produce for export? To answer this, we can examine several alternatives for OPEC behaviour. Two extreme scenarios suggest themselves. One extreme possibility is that OPEC will increase output in order to main-tain a constant price. A second extreme is that OPEC will not increase output but will let prices increase in accordance with market forces. Although these two scenarios are extremes, their implication for oil revenues may be indicative of OPEC's possible behaviour.

Table 3.3 gives the necessary information to assess possible OPEC behaviour and prices.

If we assume an average income growth of 4 per cent in the world economy over the next six years, what would this mean in terms of in-creased oil demand at constant prices? Using an average long-term in-come elasticity of roughly 1, the necessary increase, in oil exports to maintain a constant price, is given in the first row of Table 3.3.[1]

As an extreme case scenario, that is with high annual world growth of 4 per cent, no reduction in elasticities and no extra output outside of OPEC, it can be assumed that all of this increased demand could come from OPEC sources. If an equivalent output of oil was forthcoming, prices would stay roughly constant. In the second row, the implied extra revenues to OPEC are calculated, assuming a constant price of $30/barrel. In row three a very different calculation is done. That is, what would happen if OPEC did not increase output but let prices rise as dictated by increased demand. Using a price elasticity of –0.35, the implied price of oil with constant OPEC exports is calculated.[2] And in row four the implied extra revenues to OPEC are calculated for this case.

The question is, therefore, whether OPEC will choose the first option or the second.

Clearly up to and through 1986, OPEC revenues would be higher if output was increased, rather than prices. After 1986, the path of prices

Table 3.3: Oil Production and Revenues Under Two Alternative Scenarios, 1984–1990

	1984	1985	1986	1987	1988	1989	1990
1st Scenario: Constant Price with Increasing Output							
Increase in world oil demand, 1983 (MBD)	2.1	4.3	6.7	9.1	11.5	14.1	16.86
Increase in OPEC oil revenues, 1983 production at constant price ($30/barrel; million $)	23,490	47,640	72,900	99,150	126,450	154,860	184,410
2nd Scenario: Constant Output with Increasing Prices							
Oil prices at constant level of OPEC exports ($/barrel)	33.60	37.60	42.15	47.21	52.87	59.21	66.32
Increase in oil revenues at constant of 1983 exports resulting from price increases of 12% annually (million $)	21,240	44,384	70,956	100,506	133,561	170,586	212,109

would depend on whether certain members of OPEC would benefit more by increasing output or by letting prices rise. Given the high levels of excess capacity in some OPEC countries, the future of oil prices depends on two things: the relative benefits of the two scenarios for high excess capacity countries, and the ability of those countries to convince the other producers (members and non-members of OPEC) not to upset the market. Taking rough production figures available for 1984, one can see that approximately 15 MBD, or over 90 per cent of the extra capacity, is in seven OPEC countries — Iran, Iraq, Kuwait, Libya, Nigeria, Saudi Arabia and the United Arab Emirates; while a little over 60 per cent of current OPEC exports derives from these same countries. The issue then becomes, is it to the advantage of these countries to get 90 per cent of row 2 or 60 per cent of row 4? Given the low level of exports of the excess capacity countries, resulting in a total share of only 60 per cent of OPEC exports, they would be better off in every year by getting 90 per cent of row 2, as opposed to 60 per cent of row 4. In the recent past the reverse of this situation existed, that is the financial incentive was to let prices increase. But, as in both time periods, the market price is determined by the marginal producer.

As a result of these simple calculations one could conclude that it is in the interest of OPEC excess capacity producers to maximise their medium-term revenues. The question then becomes: will they pursue this course of action and will the market behave in their favour? If the answer is yes, then it is possible that oil prices may not increase before 1990. To throw more light on this matter, two important points are worth emphasising.

First, most OPEC countries, especially the so-called surplus countries, are heavily dependent on oil export revenues. On average, the OPEC-members of the GCC member states derive, from oil and oil products, over 95 per cent of their export revenues. In other words, these countries currently enjoy very little export diversification. As a result, they are extremely vulnerable to conditions in the world oil market. Their need for reserves, as indicated by reserves to import ratios, is, therefore, much higher than that for more diversified LDCs. In essence, foreign assets, in the medium term, offer a source of income diversification. This additional source of income is invaluable when the oil market is soft. As a result, if these countries have an opportunity to accumulate reserve assets, as indicated by an increase in oil output versus a price increase, they might take the former option.

Second, many of the excess capacity countries would have close to 50 years of oil output even if they produced at the capacity rate of production. As a result, medium-term considerations take on increased importance.

It is, therefore, possible to conclude, on economic grounds, that oil prices may not increase prior to 1990, given current market conditions and expectations. Clearly, unforeseen disruptions could alter such predictions. Beyond 1990, however, prices could start to rise, especially if production outside of OPEC does not increase and/or consumption growth does not decline.

These same results are obtained from a more 'sophisticated' world model.[3] The results of the model indicate that the equilibrium will remain at its present level until the end of the decade. At this price, the demand for OPEC oil will reach 27 MBD if the non-OPEC supply of oil is not increased. Under such circumstances, OPEC's oil revenues will increase by $143 billion by 1990. However, if OPEC limits the level of production and lets the price go up, its annual revenue from the export of oil will increase by about $100 billion by the end of the decade. This result is somewhat different from what was obtained through our 'back of the envelope' calculations in Table 3.3. The difference is due to the fact that our world model takes account of the supply elasticity of other sources of energy. Therefore, in the scenario where the price of oil increases, the supply of other sources of energy will be increased, preventing the price from going up at a very sharp rate.

The results of the model provide an even stronger support for our suggestion that OPEC is better off to increase production, at a constant price, rather than to increase price by restricting output to its current level. It should be noted that the results of the model indicate that OPEC as a whole, as opposed to only excess-capacity countries, is better off by increasing output instead of letting prices increase. This is because with the world model, revenues over the period, for OPEC as a whole, are higher if output is increased; whereas in the 'back of the envelope' calculation, revenues over the period are not higher for OPEC but are only higher for excess-capacity countries.

Finally, unless there are unforeseen disruptions, there are indications that oil prices will not increase before 1986 and may even stay where they are until 1990. If this conclusion proves to be correct, what are the other policy implications?

(1) For OPEC, a new hard look at the ideas to be discussed in the long-range strategy is called for.
(2) For consuming countries, imposing additional taxes on imported oil will have an effect similar to a price rise. That will only play

into the hands of those who do not want the market to take its course.

Notes

1. That is a 4 per cent increase in income will result in a 4 per cent increase in the demand for oil.

2. That is a 4 per cent increase in income would result in a 4 per cent increase in demand which could be nullified by a 12 per cent (4/0.35) increase in price.

3. In this model the world-wide demand and supply of energy, broken down into six regions, are estimated. On the demand side, the model calculates the total energy consumption of each region in each economic sector (industrial, transportation, commercial and residential). On the supply side, the model computes the domestic supply of oil, coal, natural gas and primary electricity. The difference between the total energy demand and the (domestic and imported) supplies of other types of energy determines the demand for oil. The demand and supply of oil are then aggregated to the world-wide level where they interact to determine the equilibrium price.

IV

THE SOCIAL IMPACT
OF HIGHER OIL REVENUES
ON THE GULF REGION

Dr Mohammed Al-Rumaihi,
Editor-in-Chief of *Al-Araby,*
Kuwait

IV

THE SOCIAL IMPACT OF HIGHER OIL REVENUES ON THE GULF REGION

I hope that in my talk today I will be able to draw for you as accurate a picture as possible of the contemporary Gulf States, both as they are today and as they hope to be in the future. Looking to our recent past, it is true that nomadism, tribalism and subsistence economy prevailed until the past three or four decades. But this society was not devoid of civilisation and culture. On the contrary, we had a rich and diverse civilisation, shared with our fellow Arabs and Muslims. Seafaring, trading, agriculture and nomadism were the mainstay of the economy, and formed the basis of a whole culture. The region's links with the outside world did not begin with oil — on the contrary, people from the Gulf have a long tradition of trading in East Africa, on the Indian subcontinent and right up to South China. Their dhows may have been humble, but they had a wealth of navigational skills.

It should not be forgotten that the Gulf States started to introduce modern education and health services and to update their administration well before the dawn of the oil era. Building on the basis of the Koranic school system, Kuwait and Bahrain started to introduce modern education as early as the 1920s, and publishing and newspaper production started around the same time. People were thus already prepared for the great wave of economic and social development made possible by oil revenues.

The oil states have channelled a substantial proportion of their oil revenues into building up a welfare state to enable all citizens to benefit from their national wealth. The welfare system includes free education and health, cheap power, water and fuel, free or subsidised housing and a host of incentives to promote the citizens' economic activities.

While the Gulf States have launched massive development programmes on the basis of their oil revenues and have undertaken some of the world's most ambitious projects in infrastructure, hydrocarbons, industry and agriculture, the planners of the area are continually aware of the need to invest not only in projects but in the development of human resources.

The priority given to human resource development is shown by the rapid growth of the educational system: I like to refer to this as the Gulf's 'silent revolution'.

Just 15 years ago there wasn't a single university in the area. Today, we have nine universities with two more in the making. Only a few days ago the foundation stone of the latest university was laid — The Gulf University in Bahrain. Two features make this institution unique. First, it is a joint project of all the Gulf Co-operation Council (GCC) states plus Iraq, and secondly its syllabus is devoted to those subjects in which the Gulf urgently needs qualified nationals — science, technology and medicine.

To give you an indication of the success of the educational development programmes in our area, 95 per cent of Bahrain and Kuwaiti children at primary and secondary school age are in school — a remarkably high proportion by the standards of the developing world. Some of the 5 per cent who do not attend suffer from physical or mental handicaps, but others are kept from school by social constraints — in certain sectors of society for example, school attendance by girls is still frowned upon.

Another area of the welfare state, health, has made great strides in recent years. According to the World Bank, there was in Kuwait one doctor for every 590 people by 1980, a great improvement from 1960, when the ratio was one doctor to 1,210 people. Life expectancy had risen to 70 years in 1981 compared with 60 years two decades earlier, while infant mortality dropped from 89 to 33 per 1,000.

A generous retirement scheme for all working people was introduced in Kuwait in late 1976 which takes 5 per cent of the salary from the employee and 10 per cent from the employer. The government puts in an equivalent sum.

The programme of subsidies in the Gulf is designed not only to alleviate the economic burdens of rapid development on residents, but also to stimulate their participation in the development process. In recent years legislation has been introduced to boost the role of indigenous companies, including contractors in development projects. The next Saudi five-year development plan for 1986–90 envisages a much greater role than ever before for the private sector, as long as proposed projects are shown to be economically viable.

To be frank with you, the enormous pace of economic development in the Gulf has not been without social cost. One of the greatest challenges currently facing Gulf societies is the huge number of non-national workers in the labour force. Aside from Iraq, the Gulf countries, under the enormous pressure to complete their massive development schemes quickly, were forced to bring in hundreds of thousands of expatriates, ranging from top-level managers to unskilled workers, from scattered regions of the world.

A major problem for the Gulf States is that of keeping their cultural identity intact in the face of a variety of imported influences, and of maintaining their political stability. The situation is particularly critical in the United Arab Emirates, where it is estimated that only about 27 per cent of the population is comprised of nationals. Nevertheless, a fair proportion of the expatriates are Arabs, who do not pose a direct threat to the local culture. Gulf governments are making efforts to reach agreements with the Arab labour-exporting states, such as the Maghreb countries of Tunisia and Morocco, to make sure that when expatriates must be employed, an increasing proportion are Arab.

The formation of the Gulf Co-operation Council (GCC) has been one of the most significant events to occur in the area in recent years. The council was formed in 1981 by six states who felt that they shared a common destiny and similar political and cultural milieu, and faced a common security threat. The GCC's legal status is unique. The council's secretary general, Abdullah Bishara, describes it as similar to that of a confederation, which will lead one day to unity.

Co-operation between the Gulf States is gaining momentum in a number of spheres — political, economic and cultural. The Unified Economic Agreement, which was approved at the GCC summit in Riyadh in November 1981, is the backbone of this co-operation. The agreement stipulates, among other things, that the Gulf nationals should be treated equally throughout the region, with total freedom of movement and an equal right to work in any GCC country. The GCC is aiming at a Gulf common market. As a prelude, there is free movement of capital and goods, and freedom to own property in all constituent states.

The GCC is planning a number of joint ventures, and has created the Gulf Investment Corporation with a capital of $2.1 billion. Among the joint projects under consideration or at the planning stage are the Gulf railway, a new pipeline and refinery in Oman, a Gulf gas grid and a strategic food reserve, possibly located in Fujeirah. One piece of construction which will increase Gulf economic interaction is the Bahrain-Saudi causeway now being built.

The members of the GCC are drawing up joint policies to try to tackle some of their common problems, including the social dilemmas generated by the pace of development. They are formulating a joint manpower policy. Taking into consideration the large number of young nationals due to enter the labour market in the near future, the national workforce will increase from 1.8 million in 1980 to 6.5 million in the year 2000. Especially since there has been a tendency to neglect vocational training in the past, it is vital that education and training are directed towards providing the area's youth with the skills and qualifications needed to take over key roles in the economy in the future.

The GCC is also co-ordinating moves to rationalise expenditure both on development projects and social welfare. Subsidies on fuel, electricity and water, for example, are being reduced.

Despite having been able to dwell only briefly on the major issues in the Gulf today in this talk, I hope I have succeeded in conveying an outline of current developments in the area. Although it is true that a little knowledge can be a dangerous thing, I hope that my remarks, brief as they were, will provoke some of you here to become more familiar with my homeland.

DISCUSSION

After the speakers finished, a number of questions were posed and observations made. Among them were those summarised below.

(1) Some audience members felt that too much attention is paid to the effects of supply side shocks on the industrialised countries, and too little on their effects on the developing countries who are certainly less well equipped to deal with those shocks. Furthermore, in developing countries, the main effect of supply side shocks is not on consumption of energy, but on levels of output. By adversely affecting production, energy supply shocks also adversely affect developing countries' capacity to service external debt. It was observed further that the industrialised countries have the IEA to help cushion them against oil supply shocks, while the developing countries have no such mechanism.

(2) Several members of the audience commented on the use of energy coefficients in the techniques used in estimating world demand by some of the speakers. It was pointed out that the current recession had lasted longer than normal; and that it was associated with a contraction of the energy-intensive sectors of the economy — iron and steel, automobiles, etc. The result is that statistical calculations of energy coefficients are likely to produce distorted results. Because of the underestimation of the energy coefficients, any sudden increase in demand would put a degree of upward pressure on oil prices that the estimation techniques employed would not be able to predict. In a similar vein it was noted that many developing countries are now engaged in a process of energy-intensive industrialisation that most estimation models do not adequately take into account. Once again the result is downward bias in the energy coefficients.

(3) The speakers countered criticisms of the use of the energy coefficients by contending that since these coefficients were calculated over

59

the 1979–83 period, the time frame was long enough to obviate serious underestimation. The coefficients being calculated would reflect more than simply one or two phases of a business cycle. They doubted that industrialised countries would have any real incentive to go back to less energy-efficient production techniques.

(4) Several members of the audience pointed out that all estimation techniques employed by the speakers took the political situation as given. When Dr Abdullah El-Kuwaiz was asked how vulnerable the price forecasts were to political changes, he replied 'the numbers are interesting, they are comforting, but I do not think they mean very much'.

(5) On the question of politically induced supply shocks, Dr Atkinson insisted that the West's capacity to ride them out had been greatly increased in recent years by stockpiling and diversification of sources. In his opinion, the biggest threat to the accuracy of the estimates being made by various bodies is not the possibility of a supply shock but the disruption of world economic growth. Higher interest rates and increases in protectionism could stifle the growth of world trade, and throw off the viability of demand estimates. Sagging demand would put yet more downward pressure on prices.

Dr Atkinson also pointed out that short run supply disruptions could be weathered easily, not only because of stockpiling in consumer countries, and not only because of the amount of floating storage capacity currently in use, but also because some OPEC countries can quickly increase production to offset the effects of supply interruptions in others.

V

PETROLEUM POLICIES OF THE ARAB COUNTRIES AND THEIR RELATIONS WITH THE INDUSTRIALISED COUNTRIES*

Dr Abdelkader Maachou,
Adviser to OAPEC,
Professor at the University of Paris
(Paris-Sud), France

*Translated by Peter Dewhirst

Politicians, researchers, economists and members of Canadian and Arab universities are meeting here, under the aegis of the Committee for Planning and Co-ordination of Petroleum Information (CPCPI) and of McMaster University at a difficult conjuncture of world events. As a result, this meeting has a special meaning.

It illustrates a form of interdependence between petroleum-producing countries belonging to the two main groups of countries — the developing countries and the industrialised countries.

It also illustrates responsibilities which go beyond the petroleum-producers and exporters, and beyond questions of the moment, responsibilities which relate to relations between man and his neighbours, beyond the seas and across the continents. Responsibilities and questions for men facing up to the problems which characterise this end-of-century. In other words, co-operation for development and growth on the one hand, and on the other hand, an interdependence which is not inconsistent with independence.

This encounter corresponds, therefore, to a determination stemming from a mutual recognition of the new situation which has set its seal indelibly on the international relations between the petroleum-producing countries and the industrialised countries, particularly those in the Western world.

From this viewpoint, the policies of the Arab petroleum- producers and exporters, both in the OPEC and OAPEC groups and, more recently, in the Gulf Co-operation Council, are clear to the extent that underlying them is a deep determination for dialogue and co-operation.

And I must say again that this symposium responds to a need. For it is evident that, in the succession of meetings organised, particularly since 1974, across the world, between the developing countries and the industrialised countries — whether we talk about the so-called North-South Dialogue, or the Euro-Arab Dialogue or, more recently, the

63

South-South Dialogue — in all these meetings, one dialogue seems to me to have been missing. This other dialogue is that of the petroleum-producing countries, whether or not they are members of OPEC, and whether they are developing countries or industrialised countries.

In the Western hemisphere, Canada represents in a wide sense the attentive interlocutor between the industrialised petroleum-producing countries and the industrialised non-producing countries.

I would like to break down my next comments under two headings. The first is — petroleum policies of the Arab countries: the second is — reflections on future co-operation.

With respect to the first, I would say that despite the diversity of political systems which govern them, and despite their different levels of productive capacity and revenue, the Arab petroleum-producing countries have managed to work out a certain number of common principles for dealing with basic problems involving petroleum.

These principles constitute the foundations upon which rest a policy covering questions relating their production to the world market, to prices in relation to their own growth and growth in other countries and, finally, to the contribution to international readjustment and to Official Development Aid (ODA) that the Arab producers should make.

It may seem unnecessary to mention the problems of supply in the present petroleum scene since we can no longer speak of shortages, at least for the immediate future. The current glut is a temporary situation; it has resulted in part from the commitment of Arab oil-producing countries to supply any shortfall in the world oil market. But while the Arabs remain committed to this end, the industrialised countries have used this noble commitment to further their narrow and immediate interests.

In this connection, it is worth remembering that even at the height of the oil crisis, in 1973–4, the Arab countries' petroleum policy was constantly inspired by one principal objective: meeting demand and guaranteeing supplies for the world economy.[1]

It is not necessary to dwell at length on the likely petroleum supply situation for the Western world over the next decade or two. It is clear that recovery in economic growth could lead to tensions over potential supplies, and that OPEC crude, particularly Arab crude, will play a more important role in regulating both production and prices in the near future.[2]

The decisions previously and consciously made by the Arab countries show an acute sense of responsibility, not in relation to a particular and temporary situation, but with regard to their long-term attitude towards global energy supply and global economic stability.

Thus, the decision to freeze prices for two years, after the first price adjustment, allowed the industrialised countries' economies to adapt without serious damage or difficulty. And again, in 1983, we saw a price cut and another price freeze.

It is obvious that, taken in isolation, the Arab countries have little incentive to produce beyond their own immediate needs. Furthermore, we must clearly differentiate between liquid assets and national wealth. What was sometimes improperly referred to as 'surplus' in some Arab countries — but which no longer exists to any extent today — was no more than the conversion into currency, subject to fluctuations and inflation, of some part of the Arab capital endowment. Furthermore, this capital was extracted from the soil not just to meet the producing countries' own needs, but also to take into account the supply needs of the industrialised world.

The Arab petroleum reserves constitute a form of capital subject to depletion at a rate given by the rhythm of production. Converting the production into currency does not really represent net revenue. That is why standard notions of wealth in the Arab petroleum-producing countries, linked to the concept of Gross National Product, does not correspond to economic reality.[3,4]

To underline the real dimension of the Arab countries' commitment to guarantee world supplies, we have to remember that the petroleum reserves constitute the principal — sometimes the only — source of revenue, the quasi-unique source of Arab energy supplies, today as well as in the future.

Although the Arab countries do have more than 50 per cent of world petroleum reserves and also important gas reserves, it is important to note that they possess only about 9 per cent of the planet's *recoverable* energy reserves. This figure shows up in its true light when you compare it with the US figure, which is of the order of 25 per cent, even excluding the latter country's energy resources from complex technologies like nuclear or solar power, or similar alternative sources.

This brief outline will, I hope, help to clear up a few deep-seated misunderstandings concerning Arab attitudes in general and Arab petroleum policy in particular.

This petroleum policy is based principally on three premises. The first is that cheap petroleum, notably from Arab countries, contributed in large measure, and for decades, to the prosperity of the industrialised countries. And it is still contributing today, thanks to the determination of the producing countries, who are fully committed to guaranteeing the transition to new sources of energy. Despite the successive price rises, crude oil remains today one of the cheapest forms of available energy.

By undertaking to guaranteee regular supplies to consuming countries, without making use of dubious cartel practices, the producing countries are taking a grave gamble on the future. Energy consumption of OPEC member-countries, which is today globally some three MBD, is expected to rise gradually to nearly 17 MBD by the year 2000. It would then be the equivalent of almost the whole of OPEC's production today (about 17.5 MBD).

There is no denying that, although the world is going through a transition period in energy supplies, hydrocarbons remain a major component of the overall energy scene.

The second premise underlying Arab petroleum policy is that the producing countries base their pricing decisions on the world economic situation. They have not hesitated, despite the cost, to freeze crude oil prices when the world economic situation required it. And in addition to the freeze, which is in effect a cut in the *real* value of petroleum, the price of reference crude was cut by $5 a barrel in 1983.

The final premise on which Arab petroleum policy is based is that one of the essential objectives of the producing countries is price stability. Prices should settle down at a level which will allow a *correct* compensation to the exporting countries, who have not given way to the temptation to snatch quick profits during the period when there was a seller's market.

Yet faced with this Arab petroleum policy, the policy of the major consuming countries seems to follow principally two objectives: *one*, reduce dependence on OPEC petroleum; and *two*, build up petroleum stocks which can then be used as levers to shift the market and, therefore, to alter the balance of forces in favour of the industrialised countries.

Of course, faced with this state of affairs, it is always possible to invoke questionable theories, to explain away or to try to justify certain stands. With the crisis going strong, there are plenty of these. But the crisis spares no country. It certainly leans more heavily on the developing countries, including those who are petroleum-producers. That is why we should steer resolutely towards the formulation of a framework for balanced and mutually profitable co-operation.

I will turn now to my comments under the second heading — the prospects for future co-operation.

The possibilities for co-operation can only be actualised by reconciling the urgent needs of the developing countries, and some of the demands of the economies of the industrialised countries.

Obviously, co-operation presupposes a political determination in practice. This must show itself in tangible acts meeting the legitimate concerns of the partners in question. And this in turn demands a frank discussion, without pre-set demands, which will help clear the ground and blaze the trail for the future. But a number of pertinent points should be noted, which, though they should be obvious, are too often ignored.

The Arab countries, and notably those in the Gulf region, find themselves today, like all the Third World countries, confronted by the serious problems of development. These are aggravated, on the one hand, by the state of war prevailing in this part of the world for the past four years and, on the other hand, by the permanent threat of destabilisation looming over the whole region due to the bellicose and expansionist policy of Israel.

This means the colossal efforts expended by the Arab countries to develop are thwarted by the obligation to mobilise precious resources in order to meet essential security and defence needs.

Linked to this situation — which has been in existence practically since the end of World War II — and running parallel to it, the effective life of overall Arab petroleum reserves is estimated at 41 years (on the basis of the 1979 average production) and up to 72 years (on the basis of 1982 average production). [5]

A period of 41 years, or even 72 years, is not normally important in the life of nations. It is, however, decisive in the case of the Arab petroleum-producing countries; their long-term future depends on what happens inside that limited time period.

For the producing countries their petroleum revenues are crucial to develop their economies and to accelerate the process of diversification of their economic structures. This diversification, if successful, will be able to generate fresh sources of revenue as an offset against depletion of their present energy reserves. This would allow them to free themselves progressively from the shackles of petroleum.

The consuming countries, on the other hand, are still locked into the need to import this raw material which is so indispensable to their economies and so essential to their industrial growth.

For them, petroleum is the surest and the least costly energy source to use to get them through the period of transition, a period of transition in which they need to limit its impact on economic and social life, until they can achieve a greater level of energy self-sufficiency.

Thus, it would seem that development and its needs, on the one hand, and economic growth and its demands, on the other hand, might well coalesce. This would permit a healthy and fruitful co-operation between the Arab petroleum-producing countries and the industrialised countries, including Canada.

It is not possible to tackle the problems of co-operation without placing them in the context of the current economic crisis, which is inevitably used by the industrial countries each time there is any mention of financial or technological co-operation or of opening up markets.

First of all, *which crisis* are we talking about? This word 'crisis' has been invoked almost inevitably, until quite recently, to describe a deep-rooted idea. An idea which, it must be admitted, has often been maintained and cultivated at the level of public opinion both by politicians and by the media. They brought it back each time, sometimes more or less in undertones, to the level of what it had been agreed would be referred to as the 'energy crisis'. This meant the situation sparked off in 1973, following the re-adjustment of those petroleum prices which had remained blocked solid for nearly a quarter of a century.

Today, when we talk about 'the crisis', we can say quite flatly it is no longer merely a crisis of energy but a multifaceted crisis.[6]

Viewed from the perspective of industrialised countries, it shows itself as a crisis of unemployment, inflation, industrial redeployment, technological change and international debt. The indebtedness of the developing countries has now reached the incredible figure of US$800 billion, sufficient to cause general alarm for the whole international financial community.

Looking through this prism, we see that when petroleum prices were re-adjusted in 1973 and 1979, they provoked a slowdown in world energy consumption. The growth rate of energy consumption, which had risen by 50 per cent during the previous decade, remained no higher than 10 per cent between 1973 and 1982. In addition, overall consumption dropped by 9 per cent *per capita*, for various reasons which are so well known we need not go into them here.

It is now recognised that the industrialised countries' crisis resulted from technological changes linked to structural industrial changes which call for varying periods of adaptation and adjustment, according to the country in question. The crisis hitting the Western countries did not, however, spare the Communist bloc countries and even less the developing countries.

Viewing this from the petroleum-producing countries' angle, the crisis shows up as having caused them a grave injury because of the drastic reduction in their exports and, therefore, in their revenues. The negative effects which accompanied the drop in demand for petroleum from OPEC member-countries in general, and from the Arab producers in particular, provoked a reversal of the market tendency and produced a buyer's market. [7]

But this situation has also a whiplash effect on the industrialised countries' own economies. Development programmes of the Arab countries were revised downwards as a result of the decline in oil revenues. Furthermore, the important assistance given by the Arab petroleum-producing countries, particularly those of the Gulf, to the developing countries was also reduced substantially.

A major part of this aid was used to finance goods and services coming from the industrialised countries, the very opposite of assistance granted by these industrialised countries, usually in the form of export credits or, even if the aid is not tied, for payment of goods and services supplied by the donor country.

The Arab countries have granted an appreciable amount of aid for development, both bilaterally and multilaterally. These countries have become important donors of ODA and they have thus brought a new dimension to international relations. [8]

The movement of restructuring and adjustment now under way in the industrialised countries is only just beginning, according to highly qualified specialists. And it is clear the movement does *not* have its origins in the events of 1973, for what is commonly called the 'energy crisis' had been prepared and preceded by elements which made it a revelation (an outcome) rather than a determining factor.

May I remind you that the 'energy crisis' had been preceded by a 'monetary crisis'? In August 1971, President Nixon had decided to suspend the convertibility of the US dollar against gold, thus opening up the road to its devaluation. All other currencies went 'floating' from then on. The Bretton Woods system established in 1944 had practically come to an end.

Some studies show that the forerunners of the energy crisis were several signals that were spotted in the drop in productivity, at the end of the 1960s and the start of the 1970s. The growth rate of the world economy was already beginning to run out of steam. [9]

However, the crisis could also be a factor in increasing interdependence, and it is true enough to say that the road to co-operation and interdependence is not always an easy one. In the minds of people in the industrialised countries, it leads inevitably to the creation of competing industrial structures in the developing countries, to the acquisition of technological know-how and to the opening up of markets to products from the Third World. All of which raise very delicate problems.

For the producing countries, that is, the developing countries, this road leads to a finer perception of the problems of economic growth in the industrialised countries which are, to some extent, linked to energy production and to energy prices. The petroleum-producing countries have, as we have seen, shown themselves to be attentive to these problems.

All these factors merely served, in fact, to throw light on some problems studding the difficult road to co-operation. They may have contributed perhaps to illuminate the debate on how to work out a framework for co-operation in economic, scientific and cultural activities and other endeavours.

These sectors could include:

— exchanges in the field of research and development of new sources of energy and, in particular, the exploration of oil shale, a field where Canada has acquired practical experience;
— the development of optimum working conditions for existing petroleum reserves; and
— co-operation which also takes into account the determination of the Arab countries to consolidate the foundations of their petroleum industry by upgrading all or part of their production.

To conclude, let me underline the fact that the Arab countries hold, in addition to many other material and human resources, most of the exportable petroleum and gas reserves of the world. These resources could continue to facilitate the industrialised countries' prosperity but only if the Arabs are compensated fairly and in real terms. The Arab world's extensive economic potential could work for themselves and for the good of the world through effective co-operation.

Canada, an industrialised country which is a hydrocarbon producer well versed in the problems and hazards of the petroleum industry, is well placed to discern with sympathy and understanding the anxieties of the Arab petroleum-producing countries.

It could contribute eminently to the working out and development of a model of co-operation which could serve as an example to other industrialised countries.

This, without any doubt, is the target that we could propose at this symposium.

Notes

1. Presented by Henri Simonet, vice-president of the Commission of the European Community, at Versailles in November 1975 (OAPEC/IFP/Paris colloquium).

2. See *Arab Oil and Gas*, no. 363, 1 May 1984: Studies and Documents, a tentative synthesis made by IFP at the Second Saudi Petroleum Exhibition, El Khobar, April 1984. See also, *Market Monitoring Bulletin* and forecasts by the Enerfinance Company, in Paris.

3. Arab Monetary Fund.

4. International Monetary Fund.

5. Report of the Secretary-General of OAPEC, 1982.

6. Edgar Morin, quoted in 'Ten years for a prologue' by Philippe Lefournier, special number of *L'Expansion* (no. 223).

7. Dr Ibrahim Shihata, 'Another face to OPEC'.

8. Ibid.

9. See 'Ten years for a prologue', special number of *L'Expansion*.

VI

CANADA'S NATIONAL ENERGY POLICY

Dr P.M. Tellier,
Deputy Minister of Energy,
Mines and Resources, Canada

I am very pleased to be participating in this symposium. It serves as a very useful forum for an exchange of information between oil- producing and consuming nations. Such an exchange is essential in view of the economic interdependence of both groups of nations, especially given that the energy policies of oil-importing countries are influenced strongly by international developments. It is my intention, today, to review briefly the evolution of Canadian energy policy and to outline some of the key issues which Canada will be facing in the future.

Canadian energy policy has evolved considerably over the years — often in response to world oil supply and price conditions. As in other countries, the oil price shock of 1973/4 had a major impact on energy pricing and trade policy. Similarly, the 1980 National Energy Program (NEP) was, to a great extent, a response to continued turmoil in the international energy environment.

At that time, world oil supplies had once again taken a dramatic leap in price and had become very uncertain. This was worrisome because we were importing over 400,000 barrels of oil a day (B/D), representing about one-quarter of our consumption. Furthermore, unless action was taken, our appetite for imports was expected to increase by 50 per cent by 1985. The government of Canada believed that our import dependency was unnecessarily high, given our domestic oil supply opportunities, and existing surpluses of natural gas and electricity.

Therefore, the NEP made *energy security*, including oil self-sufficiency, one of its key objectives. Its other major objectives were: *opportunity* for Canadians to participate in the energy industry and to benefit from employment and business opportunities related to the expected growth in the energy sector: and *fairness* in consumer oil prices and the distribution of energy revenues.

These three objectives were, of course, not new to Canadian energy policy. However, the NEP represented a watershed in our energy policy

75

in that it *established a comprehensive set of initiatives to achieve these objectives*. Initiatives included demand management programmes, favourably priced natural gas, the blended oil price system, the Petroleum Incentives Program, an increased role for Petro-Canada and a new Canada Lands fiscal and management regime. While the objectives were widely accepted in Canada, some of the above instruments were controversial, especially those dealing with taxation, pricing and Canadianisation. Much of the controversy was caused by the lack of an energy agreement between the government of Canada and the energy-producing provinces. This controversy was substantially moderated with the Canada/Alberta deal in 1981.

The 1981 energy agreements with the producing provinces resulted in significant changes to NEP programmes, taxes and pricing policies. In the three years since the 1981 agreements, the NEP has evolved still more, as many of the assumptions upon which it and the energy agreements were based, such as higher world oil prices and strong economic growth, did not materialise. Thus, the 1982 *NEP Update* responded to industry cash flow problems, just as the 1983 amendments to federal-provincial energy agreements assisted industry to cope with a situation of declining world oil prices.

Thus, just as the changing world oil market shaped Canadian energy policy in the early 1970s, and in 1980 when the NEP was first introduced, federal energy policy has evolved considerably since 1981 in response to the recent softening of world oil markets, and other factors. Changes have, however, been consistent with the key objectives of the NEP. Consequently, we have made very considerable progress towards the NEP objectives — objectives which still enjoy broad support in Canada.

Higher consumer prices, together with NEP demand management programmes, led to a 22 per cent reduction in oil demand between 1980 and 1983. This, of course, has greatly enhanced our *energy security*, not to mention our energy trade balance. Canada was a net exporter of oil in 1983 and is expected to remain so in 1984. While substantial new sources of production will be required to maintain self-sufficiency, *Canada does not now have the oil problem we faced in 1980*.

Considerable progress has also been made in providing an *opportunity* for increased involvement by Canadians and Canadian firms in energy resource development. There has been a major restructuring of the industry, as Canadian ownership of petroleum production revenue increased from 28 per cent in 1980 to 38 per cent in 1983.

There have also been numerous farm-in arrangements in the Canada Lands being promoted in part by the Petroleum Incentives Program and our Canada Lands management regime. As a result, Canadian companies have an opportunity to earn an interest in 62 per cent of the Canada Lands, as compared to only 38 per cent in 1980. Furthermore, the number of Canadian companies operating drilling rigs on the Canada Lands increased from three in 1980 to seven in 1983.

Progress has also been made in ensuring *fairness* in the distribution of burdens and benefits of energy resource development. The federal government's share of petroleum revenue is now 17 per cent, somewhat over half of the share of the producing provinces. This is a considerable improvement from the pre-NEP regime. However, it is substantially lower than the federal take envisaged in the 1981 agreement, reflecting the impact of lower world oil prices and significant federal concessions to the industry since 1981. Industry revenue levels are up considerably from the pre-NEP regime, and even their revenue share has increased to 53 per cent of net revenues, from 42 per cent in 1979.

Consumer oil prices now stand at about 92 per cent of world levels. This parity is much higher than envisaged in the NEP, reflecting the federal-provincial pricing agreements and concessions for the industry in the light of lower world oil prices. However, while consumer oil prices have risen, increases have been much more gradual than in other industrial countries. As well, consumers have had access to a vast array of direct programmes to enable them to lower their energy costs through conservation measures and switching to less costly sources like natural gas, which is now priced at 65 per cent of oil, as opposed to the pre-NEP 85 per cent parity.

One of the fundamental pressures for change in our pricing and fiscal programmes over the past few years related to the strong public interest in ensuring the financial health of the oil industry. It is fundamentally important to Canada's energy objectives that our industry has the investment incentives, and financial wherewithal, to marshal aggressive oil development plans. There is considerable evidence that the industry is now financially capable of undertaking the investments needed for Canadian oil self-sufficiency. For example, there has been a dramatic escalation in Canada Lands oil and gas drilling activity. According to *Oilweek* magazine, well completions totalled 66 in 1983, compared to only 26 in 1980. Moreover, exploration expenditures, at some $1,950 million, were more than double 1980 expenditures of $837 million, and development expenditures were also at record levels ($130 million).

While the federal government is encouraged by high levels of offshore drilling activity, I would emphasise that it does not promote activity in any one region at the expense of others. The facts on oil industry activity in western Canada show that it does not.

— Oil drilling activity in 1983 was at record levels — some 53 per cent more wells were completed, compared to the previous record year, 1980, and 78 per cent more than 1982.
— In terms of metres drilled, 1983 oil activity outpaced 1980 by 68 per cent.
— The value of land sales rose by 46 per cent in 1983 to $565 million.
— Much of the increase in oil drilling is related to lower risk development, as opposed to exploration activity. However, the ratio of exploration to development drilling is much higher in Canada (20 per cent) than in the USA (8 per cent).
— As a result, Alberta's oil reserves increased for the first time since 1969, and many expect this to continue.

It is clear that government incentives have also been attractive to industry in the area of non-conventional oil activity. There has been promising progress on the oil sands front, where emphasis has been placed on small-scale development.

This 'phased-development' approach is much more consistent with development possibilities under the current uncertain world oil price outlook. Three companies have already taken advantage of federal/ provincial fiscal incentives: Esso at Cold Lake, BP and Petro-Canada jointly at Wolf Lake, and Amoco at Elk Point. While each of these projects entail production of around 10,000 B/D, major expansions are being considered — Esso hopes to be producing 56,000 B/D by 1988 — and other projects are on the drawing board. In addition to these bitumen projects, Petro-Canada is considering an integrated oil sands project, about half the size of Alsands, with start-up envisaged for mid-1991.

Oil companies are also responding favourably to our enhanced oil recovery investment incentives. With the availability of world-level prices, earned depletion and a reduced PGRT as well as PIP grants, this type of activity represents one of the most attractive investment opportunities in Canada and the world.

Progress has also been achieved in investment in oil-upgrading facilities. Last year, the government of Canada and Saskatchewan agreed, with the Consumer's Co-operative Refineries Ltd, to enter into the first phase of a $600 million, 50,000 B/D heavy oil upgrader project in Regina. This project, and others which are being considered, are important to Canada, not only in terms of employment and business spin-offs, but also because they would ensure the greater domestic utilisation of our heavy oil resources.

Based upon progress towards the NEP objectives, I believe that our policies have been quite successful. None the less, I am not asserting that the NEP has been without problems — problems have existed and anomalies identified. However, I would argue that an evaluation of energy policy should focus on its flexibility in responding to a changing environment. The evidence is that Canadian energy policy has been flexible, relying on close consultation with industry and other governments.

This process of policy evolution through consensus will continue as we deal with emerging energy problems: for example, in relation to natural gas. As you may know, poor US gas market conditions have been a big problem for Canadian producers, and government alike. While gas exports to the United States declined by some 10 per cent in 1983, gas export revenues fell by some 17 per cent — indicating the considerable degree of US market softness, despite gas border price reductions. Moreover, although gas export volumes decreased by only some 10 per cent, actual sales totalled only 40 per cent of authorised levels — reflecting the fact that markets could not be found for recently authorised new exports, after industry had undertaken major investments to prove up additional gas reserves.

As a result, the financial health of a number of companies has been impaired. In addition, gas-related drilling, in contrast to oil-directed drilling, was way down — down 40 per cent from 1982 and 65 per cent from the 1980 record year.

Facilitating the expansion of natural gas sales represents one of the several major energy-related challenges facing the federal and provincial governments. The federal government is currently working (in a task force) with the governments of the producing provinces, and the industry, with the objective of developing a *common* approach to a *common* interest — that of increasing gas export revenues. We are willing to listen to, and act on, industry advice, just as we did last year when the volume-related incentive price was introduced for gas exports. As to domestic gas sales, the federal government is committed to maintaining the 65 per cent parity with oil prices and will continue to look for additional ways to expand sales, again in close consultation with industry and the producing provinces.

The federal government will also continue to place a high priority on ensuring that domestic oil productive capacity is utilised to the highest possible extent. We have been able to deal effectively with the shut-in oil problem since 1982, again relying on regular consultations with industry and the governments of the producing provinces. The government has gone to great lengths to reduce shut-in. For example, direct light crude oil exports to the USA were resumed last year. While it is virtually impossible to prevent small volumes of shut-in, all parties are resolved to working together to avoid significant shut-in volumes in the future.

Another issue on our plate is to implement the Nova Scotia energy deal negotiated two years ago. This should be completed soon. The challenge remains to secure a similar agreement with Newfoundland. As you may know, on 8 March 1984, the Supreme Court ruled Hibernia and all resources offshore of Newfoundland on the continental shelf to be under federal jurisdiction. This decision clarifies the situation for industry. However, we are hopeful that we will soon be able to work jointly with the government of Newfoundland to ensure maximum benefit of this activity to revitalise the Newfoundland economy.

The general conclusion that I want to make, based upon the above observations, is that the federal government has implemented energy policy flexibility in response to changing world circumstances, and has worked well with industry to further Canadian energy interests. This has been the key to our progress thus far, and will be the key to resolving current and future issues and challenges.

These future challenges are very considerable — not just in terms of energy problems but also regarding some major energy policy questions. For example, many in industry point to anomalies of our regulated oil price system, together with the fact that domestic prices are now almost at world levels, and argue for oil price deregulation. Many of these industry arguments are persuasive. Yet, there are costs to change, and the matter requires careful consideration. Further, the government of Canada faces some very difficult decisions on some of its expenditure programmes in the light of fallen federal revenue from the energy sector and growing budgetary deficits generally. These programmes have had a very important role in furthering energy objectives and easing the adjustment by consumers to higher energy prices. Yet, expenditure programmes, for both consumers and the industry, must be evaluated not just on past performance, but also on the basis of their expected need in the future.

When grappling with such difficult policy decisions, it is very encouraging to see the extent to which business is involving itself in broad-based studies of energy policy. Many oil industry associations, individual oil companies and non-petroleum business groups, as well as independent public groups and academic organisations, are now giving considered thought to the future direction for energy policy. This represents an extremely important source of expertise and advice for governments and will help ensure that further progress can be made towards the achievement of Canadian energy objectives.

VII

THE NATIONAL ENERGY PROGRAM AFTER THREE YEARS: AN ECONOMIC PERSPECTIVE*

Dr Brian L. Scarfe,
Department of Economics,
University of Alberta,
Canada

---

*Permission has been granted by the *Western Economic Review* to include this paper in these proceedings.

Introduction: The Nature of Our Current Energy Problem

Canada is fortunate in possessing substantial hydrocarbon energy resources. Indeed, we currently export appreciable quantities of natural gas to the United States. On the other hand, we also import considerable quantities of conventional crude oil to supply markets in eastern Canada. Taking all energy resources into account (including our substantial hydro-electricity generating capacity), Canada is a net exporter of energy commodities. Her Achilles heel, however, is conventional crude oil.

Although Canada is energy self-sufficient in an overall sense, she is not self-sufficient in crude oil. Indeed, about one-fifth of her annual consumption needs are filled by importing conventional crude oil from a variety of foreign sources of supply. Although the considerable resources of the tar sands, the Beaufort Sea and nearby Arctic Islands, and the Hibernia and associated fields off the east coast may eventually be developed, only a small proportion of these resources is currently being commercially extracted, and it will take an immense amount of capital investment and (in some cases) technological innovation to bring more of these expensive resources to market. In the meantime, and for the foreseeable future, our remaining stocks of conventional crude oil, which are located largely in the three westernmost provinces, will continue to experience gradual depletion.

Prior to 1973, Canadians had become accustomed to energy resources, especially crude petroleum and natural gas and, accordingly, growth in demand for primary energy surged. Although world prices began to increase markedly in late 1973, this entrenched Canadian attitude led to widespread resistance to raising domestic petroleum prices to levels more closely approximating those of the world oil market. Consequently, while other industrial nations (with the initial exception of the United States) raised their prices to world levels fairly rapidly, thereby lowering their

85

annual rate of growth of energy demand, Canada lagged behind on the pricing issue and, not surprisingly, annual energy consumption grew at a noticeably higher rate here.

Although economic logic strongly suggests that Canada should have moved fairly rapidly towards world oil prices, a variety of reasons have been employed as partial justification for keeping wellhead prices for oil and gas below the levels that market forces would otherwise dictate. The validity of at least three of these reasons is questionable. The fact that there is a heavy degree of foreign ownership in the energy sector may provide an argument for taxation of energy rents, but not for shading internal wellhead prices from world prices. The fact that energy pricing policies may have inflationary side effects does not provide a good counter-argument to the basic need to establish appropriate relative prices. Nor do reasons connected with providing the central Canadian manufacturing sector with low energy prices to enable it to obtain a competitive edge over foreign producers have much validity. Not only are fuel costs a rather small proportion of total costs in many industrial processes, but also the maintenance of artificially low prices for just one type of input is more likely to encourage the perpetuation of inefficient production technologies than to enhance overall competitiveness in relationship to foreign manufacturing industries.

The main stumbling block in obtaining a more rational and efficient energy pricing policy has been the difficulty of agreeing on an appropriate sharing of the rents obtainable from our remaining stocks of low-cost conventional crude oil and natural gas among producing firms, energy consumers and provincial and federal governments. Nevertheless, a policy of maintaining petroleum prices well below world market levels is likely to have several adverse effects. Existing supplies of conventional crude oil are likely to be consumed more rapidly due to the reduced incentive to conserve them. Shifts to alternative more abundant energy sources such as natural gas, and to new energy-conserving industrial and transportation technologies, are likely to be inhibited. Moreover, the development of more expensive sources of supply, such as deeper geological formations, oil sands and offshore deposits is less likely to occur. In addition to this, the continued reliance upon expensive (and possibly insecure) imported oil supplies may put an unnecessary strain on the balance of payments, putting downward pressure on the domestic currency in international money markets, or necessitating larger foreign borrowing and possibly higher real (or inflation-adjusted) interest rates. This reliance may also enlarge the federal government's deficit through the subsidisation of imported oil, as appears to have been the case in

Canada prior to the establishment of the petroleum compensation charge under the NEP.

With respect to natural gas supplies, we are for the next few years substantially better off, and indeed a policy of substituting natural gas for fuel oil in industrial and home heating uses probably makes sense over the present decade. A change in their relative price is an essential ingredient in shifting demand away from the commodity in short supply, and towards that with which we are more generously endowed, and for which the longer-term reserve picture looks reasonably buoyant. Nevertheless, the recent growth in our reserves of natural gas could easily turn around if reasonably favourable netbacks are not provided to our natural gas-producing industry.

The National Energy Program (NEP), launched in October 1980, had three basic objectives: security of energy supply, Canadianisation of the oil and natural gas industry, and fairness in pricing and revenue-sharing. Essentially, this programme attempted to hold down the wellhead or producer prices of conventional crude oil and natural gas in relationship to world market prices, and to increase consumer prices of these products to aid in conservation (with natural gas prices increasing less fast than oil prices to aid substitution of gas for oil). The NEP also sought to generate large-scale revenues for the federal government from the tax wedges proposed between these two sets of prices and from the petroleum and natural gas revenue tax (PGRT), initially set at 8 per cent of net production revenues, and to create incentives for further exploration and development of non-conventional hydrocarbon energy resources, particularly by Canadian firms on lands outside existing provincial boundaries.

Although the NEP was supposedly designed to ensure security of oil supply for Canada, it is clear that it represented mostly a 'demand-side' solution, for the programme involved only a minimal increase in the wellhead price of conventional crude oil. Prices to consumers were augmented substantially, however. The blended or refinery-gate price from late October 1980, until 1 October 1981, increased by $9.55/ barrel, or 51.6 per cent, from $18.50/barrel to $28.05/barrel; only $2.00/ barrel of this represented a wellhead price increase for domestic producers. The remaining $7.55/barrel represented federal government levies, which include the Petroleum Compensation Charge (PCC), designed to finance the subsidy provided to users of high-priced imported crude or oil sands production, and the $1.15/barrel Canadian Ownership Charge (COC), which assisted with financing the Petro-Canada takeovers of Petrofina Ltd and the marketing and refining operations of BP (Canada) Ltd. In consequence, the blended price of crude petroleum exceeded the wellhead price ($18.75/barrel until 1 October 1981) by $9.30/barrel. The Toronto city-gate price for natural gas also rose under the NEP, but *all* of the increase prior to the New Energy Agreement (NEA) of September 1981 accrued to the federal government as a result of its Natural Gas and Gas Liquids Tax (NGGLT), then set at 45c/thousand cubic feet, and its Canadian ownership levy of 15c/thousand cubic feet.

In addition to this, the PGRT reduced producer netbacks from the wellhead prices of both oil and natural gas. The ensuing consequences of the NEP were not surprising. Drilling rigs and skilled manpower moved to the United States and many of the rigs and crews remaining in Canada became unemployed. This substantial under-utilisation of capacity continues to exist today. Sales of oil and gas rights (licences and leases) by provincial governments fell by more than 60 per cent in 1981 compared with their 1980 levels, and continue to remain very depressed today. Although conditions in the US oil and gas industry have also been depressed, and buyer resistance in the United States to high export prices for Canadian natural gas has played an important role in the severe slowdown in industry activity in the western sedimentary basin, the NEP has probably been the most substantial single cause for this downturn.

None of this has boded well for security of future energy supply. Indeed, it is quite perplexing how the erosion of producer netbacks was supposed to achieve this. Undoubtedly, the NEP has also adversely affected the rate of economic growth in western Canada, and helped to generate both substantial unemployment and numerous commercial bankruptcies. Some important adverse effects have spilled over into the eastern Canadian manufacturing sector as new orders for industrial equipment have been reduced considerably. Higher unemployment has thus been generated elsewhere, not only for this reason but also because the rate of immigration of job-seekers into the western provinces has slowed to approximately zero.

The NEA, which covers the period from 1 September 1981 to 31 December 1986, made several important modifications to the NEP, especially to pricing policies. Basically, the NEA was an arrangement whereby the Alberta government acknowledged the federal government's ability to collect substantial new forms of tax revenues from the provincial oil and gas sector in exchange for substantially higher producer prices for oil and natural gas than those contained in the NEP. Thus, the NEA does go a considerable way to rectify the pricing problems created by the NEP. Although the first priority in any compromise solution should have been to re-establish an appropriate fiscal framework for the industry, and only after accomplishing this should one have worried about the distribution of public-sector revenues between governments, in retrospect it is pretty clear that this was not really achieved. Since the NEA was signed, substantial downwards revisions to the forecast benchmark world market price for oil have led to important modifications to our internal pricing, royalty and (to some degree) taxation arrangements. These arrangements are described in the next two sections of this paper.

Current Pricing Arrangements

Two basic categories of oil were distinguished in the NEA, namely old oil and new oil. Conventional crude oil discovered prior to 1 January 1981 is designated as old oil, whereas oil discovered after that date, or produced from the oil sands and offshore sources, or by tertiary recovery techniques from existing pools, is designated as new oil.

The federal government has conceded that the wellhead price of old oil should be increased more rapidly than under the NEP, at $4.50/barrel during 1981 and 1982. For the next four years, this figure may increase at the rate necessary to keep domestic prices at no more than 75 per cent of the actual international price of oil (or average cost of imported crude oil laid down at Montreal) less transportation costs from Alberta to Montreal, provided that this does not exceed $8.00/barrel per year (more precisely, $4.00/barrel on each six-month anniversary date).

In fact, since the international price of oil has recently fallen to about $38.00/barrel in Canadian funds (with a world f.o.b. price of about $29.00 US/barrel), the current wellhead price of $29.75/barrel set on 1 January 1983 has already breached the 75 per cent cap by approximately $3.00/barrel, after appropriate allowances for transportation costs (see Table 7.1). The 30 June 1983 Canada/Alberta Amending Agreement freezes this wellhead price at $29.75/barrel as long as it remains within the 75–100 per cent of world price band. One of the consequences of the much lower forecast for world oil prices is that all revenue-sharing and netback estimates calculated under the NEA, and to a considerably lesser degree those under the 1982 *NEP Update* must now be judged to be totally misleading.

For new oil, permissible producer wellhead prices are more generous than for old oil. Indeed, the New Oil Reference Price (NORP) closely approximates the actual international price for oil. However, no increases are permitted in the NORP that would render this price larger than the international prices less transportation costs to Montreal. In fact, the

Table 7.1: Schedule of Prices and Taxes per Barrel: Conventional (Old) Crude Oil and New Oil Reference Price (NORP)

	NEP Wellhead Oil Price	Maximum New Agreement Wellhead Oil Price	Projected Actual New Agreement Wellhead Oil Price	Estimated Pipeline Tariffs	Estimated Petroleum Compensation Charge (PCC)	Estimated Blended Oil Price	New Oil Reference Price (Estimated Wellhead)	Import Cost ($Cdn/bbl)
	(1)	(2)	(3)	(4)	(5)	(6)	(7)	(8)
1 Oct. 1981	18.75	21.25	21.25	1.55	8.15	30.95	—	43.00
1 Jan. 1982	19.75	23.50	23.50	1.58	6.30	31.38	45.92	41.33
1 July 1982	20.75	25.75	25.75	1.58	6.30	33.63	43.00	41.33
1 Jan. 1983	21.75	29.75	29.75	1.71	3.76	35.22	44.71	38.61
1 July 1983	22.75	33.75	29.75	1.71	3.76	35.22	40.86	36.20
1 Jan. 1984	25.00	37.75	29.75	1.75	3.76	35.26	39.11	36.70
1 July 1984	27.25	41.75	29.75	1.75	3.76	35.26	39.65	36.70
Average 1985	30.62	47.75	29.75	1.75	3.76	35.26	39.65	36.70
Average 1986	37.00	55.75	29.75	1.77	3.76	35.28	41.00	38.50

Notes:

1. The world f.o.b. price of oil is approximately $29.00 US/barrel (or $36.00 Cdn/barrel) at the time of writing, in 1984. It is projected to remain approximately stationary during 1983, 1984 and 1985, and then increase by about 2 per cent in real terms in 1986 using the US wholesale price index as the projected inflation factor.

2. Over the period 1981–6, the New Energy Agreement of 1 September 1981 stipulates that the Petroleum Compensation Charge (PCC) is to be set so as to leave no revenue in excess of the amount required to finance oil import compensation and the subsidy on oil qualifying for the NOPR. The current charge of $3.76/barrel seems excessive from this perspective, and ought to be reduced to about $2.80/barrel in the near future. The blended oil price given in the table includes transportation charges (pipeline tariffs) to Montreal but excludes the Canadian ownership charge of $1.15/barrel.

3. The Amendment of 30 June 1983 to the New Energy Agreement stipulates that, if the international price of oil increases, no increases will take place in the conventional wellhead price that would render this price larger than 75 per cent of the actual international price of oil (or average cost of imported crude oil laid down at Montreal) less transportation costs to Montreal. It also stipulates that should the international price decrease, the domestic price shall be adjusted so as not to exceed 100 per cent of the international price, less transport costs to Montreal. Thus the domestic wellhead price will remain at $29.75 as long as this price remains within the 75 per cent–100 per cent of world price band.

4. The Amending Agreement also allows oil under the Special Old Oil Price (SOOP) programme and qualifying infill wells to receive the NORP effective from 1 July 1983. As under the original Agreement, no increase will take place in the NORP that would render this price larger than the actual international price less transport costs to Montreal.

5. We are indebted to officials at Energy, Mines and Resources, Canada for assistance with the preparation of some of the figures contained in this table.

NORP has tracked downwards from about $43.00/barrel in early 1982 to about $37.00/barrel at present. Since it is now projected that world oil prices will remain level in nominal US dollar terms throughout 1984 and 1985 at around $29.00 US/barrel, and then rise at no more than 1 per cent *per annum* in real terms until the end of the decade, the possibilities for the NORP are similarly circumscribed if this forecast turns out to be accurate.

The reference point for pricing natural gas is now the Alberta border rather than the Toronto city-gate as under the NEP, so that transportation costs from Alberta to Toronto will not directly influence what producers receive. The natural gas price was scheduled to increase by 25c/mcf every six months commencing 1 February 1982, or by $2.50/mcf in total over the five years of the NEA (see Table 7.2). Pipeline tariffs, the federal NGGLT and the federal COC imply a Toronto city-gate price which is considerably larger than the Alberta border price. However, the NGGLT is supposed to be regulated so as to generate a Toronto city-gate price which is approximately equal to 65 per cent of the BTU equivalent price of crude oil at the Toronto refinery gate; that is, the blended price of oil, which includes the PCC and the COC. Although the NGGLT once rose to 68c/mcf, it is now projected to decline to zero with the lower profile projected for the blended price of oil. Indeed, under the Canada-Alberta Amending Agreement of 30 June 1983, the NGGLT falls to zero on 1 February 1984, and the Alberta border price increases are limited to about 16c/mcf on that date and no increase on 1 August 1984, in order to maintain the 65 per cent BTU equivalency with the blended price of oil at least until 1 February 1985.

The lower price for natural gas than for crude petroleum for an equivalent amount of energy is designed to encourage substitution of natural gas for oil by energy users, and is supported by provisions in the NEP for lump-sum subsidies for those who convert to natural gas from oil. Higher prices for both fuels will undoubtedly lead to greater conservation efforts on the part of consumers. But the supply response by producers is not so clear-cut, for two other factors must be taken into account, namely expected future prices and taxation arrangements affecting producer netbacks.

First, the domestic prices of both new and old oil are tied to the world oil price and may not exceed it, or 75 per cent of it, respectively. It is now abundantly clear that the world oil price will not augment as rapidly as hypothesised under the 1982 *NEP Update*, let alone the NEA, and may well decline further before it rises again. Even with the recovery in economic activity in the Western world, the existing high world price

Table 7.2: Schedule of Prices and Taxes per Thousand Cubic Feet Natural Gas

	NEP Pre-tax City-gate Price	New Agreement Alberta Border Price	Estimated Pipeline Tariffs	New Agreement Estimated Natural Gas Tax (NGGLT)	New Agreement Estimated Toronto City-gate Price	Fall 1983 Estimated Natural Gas Tax (NGGLT)	Fall 1983 Estimated Toronto City-gate Price	Fall 1983 Estimated Alberta Border Price
	(1)	(2)	(3)	(4)	(5)	(6)	(7)	(8)
1 Oct. 1981	2.60	1.82	.78	.45	3.05	.45	—	—
1 Feb. 1982	2.75	2.07	.87	.65	3.59	.65	—	—
1 Aug. 1982	2.90	2.32	.94	.68	3.94	.68	—	—
1 Feb. 1983	3.05	2.57	.95	.74	4.28	.48	4.00	2.57
1 Aug. 1983	3.20	2.82	1.00	.91	4.73	.16	3.96	2.82
1 Feb. 1984	—	3.07	1.00	.96	5.03	0	3.98	2.98
1 Aug. 1984	—	3.32	1.00	1.06	5.38	0	3.98	2.98
1 Feb. 1985	—	3.57	1.05	1.07	5.69	0	4.03	2.98
1 Aug. 1985	—	3.82	1.05	1.20	6.07	0	4.03	2.98
1 Feb. 1986	—	4.07	1.10	1.37	6.54	0	4.08	2.98
1 Aug. 1986	—	4.32	1.10	1.52	6.94	0	4.08	2.98

Notes:

1. The export price of gas is approximately $5.40 Cdn/mcf ($4.40 US/mcf) at the time of writing, in 1984. An incentive price of $4.05 Cdn/thousand cubic feet ($3.30 US/mcf) exists for volumes exceeding 50 per cent of existing contract volumes.

2. The Toronto city-gate (or wholesale price) includes transportation charges (pipeline tariffs) from the Alberta border to Toronto but excludes the Canadian ownership charge of 15c/mcf, which applies to natural gas as well as to oil. It should, however, be noted that the Alberta government has agreed to pay Market Development Incentive Payments out of the Alberta border price to finance capital expansion of gas distribution systems and sales promotion programmes in eastern Canada. These payments will be as high as 30 per cent of the Alberta border price for all new volumes of gas sold for industrial and home heating uses there.

3. Over the period 1981–6, the New Agreement stipulates that the natural gas and gas liquids tax (NGGLT) on domestic sales is to be set at a level generating a parity relationship between the wholesale price of natural gas at the Toronto city-gate and the average price of crude oil at the Toronto refinery gate (the blended price) of approximately 65 per cent. The maintenance of such a relationship will require that the NGGLT be set to zero by early 1984.

per barrel of oil will continue to depress overall demand for crude petroleum despite some inventory restocking demand which will soon appear. Under Saudi Arabian leadership, OPEC has found that it has to restrict supplies substantially just to maintain the current price of about $29.00 US/barrel, let alone increase it.

The OPEC nations undoubtedly also recognise that any additional hike in their price, even if accompanied by disciplined supply restrictions on the part of members, will only encourage the development of high-cost alternative supplies in the industrialised world. These include reserves found offshore from Atlantic Canada, in the Canadian Arctic, the oil sands of Alberta, oil shales in the United States and elsewhere, as well as the tertiary recovery of existing conventional oil. Indeed, the realisation that world oil prices will not increase as previously thought is unquestionably one of the main reasons why oil companies have not proceeded with the early development of resources such as the oil sands, and have restricted their other exploration and development expenditures.

The second set of factors affecting industry supply responses are the taxes being imposed and their impact upon the net returns to producers. The next section discusses these taxation issues.

Taxes on Industry Revenues

A main thrust of the NEP was to appropriate for the federal government a sizeable share of the economic rents (that is, returns above those available in alternative uses of investible funds) which are available from oil and gas production. Given the negative impact of the NEP on oil industry exploration and development activity, some may have expected that the federal-provincial energy accord (NEA) would result in lower tax rates for the industry. Yet, this definitely did not occur. Any improvement in the industry's net returns comes from higher prices, not lower taxes, under the NEA. Subsequent changes in royalty arrangements under the Alberta government's OGAP programme and in taxation arrangements under the federal government's 1982 *NEP Update* became necessary, however, when the realisation of a lower world oil price profile became self-evident.

The most important new federal tax, the Petroleum and Natural Gas Revenue Tax (PRGT), was originally set under the NEP at 8 per cent of production revenues, after deducting operating costs (but not provincial royalties). This tax was the major factor which seriously eroded oil company netbacks between 1980 and 1981, and thereby discouraged new exploration and development. The September 1981 accord did not remove or reduce this tax, but as a net result of two amendments increased it even further. On the one hand, the industry was permitted to deduct a 25 per cent resource allowance in addition to operating costs to arrive at the net figure on which the PGRT was to be levied. This is something akin to allowing the industry to deduct provincial royalties; but since the latter are substantially higher than 25 per cent (at present about 37 per cent in Alberta, for old oil), royalties are not fully deductible. On the other hand, instead of just raising the PGRT to a rate which, on the lower tax base, would have provided roughly the same revenues (that is, about $10\frac{2}{3}$ per cent), the federal government set the new rate at 16 per cent, which is equivalent to a 12 per cent rate on the original NEP tax base, or 50 per cent greater than it was before. This new tax rate applied to all old conventional oil and natural gas production, and to all new oil production as well.

In addition to the PGRT, the federal government also imposed an Incremental Oil Revenue Tax (IORT) of 50 per cent on all incremental oil revenues after deduction for related Crown royalties. Incremental revenue is defined as the difference between actual revenue received and the lower revenue which would have been received under the original NEP pricing schedule. Since this additional revenue will not be counted as income when corporate income taxes are levied, the IORT really replaces the corporate income tax on it. Unfortunately, unlike corporate income tax, it tends to discourage exploration activity and to lower the incentive to commence secondary or tertiary recovery programmes for existing oil pools. Perhaps for this reason, the IORT was suspended in the 1982 *NEP Update*, initially for one and now (in the 19 April 1983 and 15 February 1984 federal budgets) for three years effective 1 June 1982. The same revenues will not be liable for federal (and provincial) corporate income taxes, a move which is supposed to be of direct help to companies reinvesting their profits in exploration activities, and thereby not paying as much income tax in any case. Indeed, one might well conclude that the IORT should be abandoned permanently in favour of the more traditional taxation measures.

The other relevant measures contained in the 1982 *NEP Update* that were designed to improve the industry's cash flow in the 1982–3 period include:

(1) a reduction in the effective PGRT rate by one percentage point to 11 per cent for one year effective 1 June 1982, though the 12

per cent effective PGRT rate has been restored in the 19 April 1983 federal budget;

(2) a $250,000 annual tax credit as an offset against the PGRT, which is especially intended to help small producers;

(3) a reduction in the effective PGRT rate on synthetic oil production to 8 per cent from 12 per cent for the two years 1983 and 1984;

(4) an increase, effective 1 July 1982, in the wellhead price of 'new-old' oil (discovered between 1974 and 1980) to 75 per cent of the landed-in-Montreal price, a change which temporarily affected the price of about 10 per cent of our conventional oil production, but which did not affect the blended price of oil since the PCC account was already generating a considerable surplus; and

(5) an extension, as of 1 January 1983, of the NORP to other oil-recovery projects, especially incremental oil obtained from tertiary recovery projects in existence prior to 1981, experimental projects, and wells from which production has been suspended for a period of at least three years. (Further favourable adjustments in the application of the PGRT to new enhanced oil recovery projects were contained in the 19 April 1983 federal budget.)

According to EMR estimates, the revenue implications for the industry of these pricing and fiscal measures were expected to total roughly $2 billion, with about half of this accruing through the PGRT tax credit spread over five years, and about half of it accruing through all the measures combined in 1982–3 alone. The 1982 *NEP Update* argued that it represented significant up-front help to the industry's cash flow problems and that it would rekindle exploration and development activity. However, the $2 billion tax relief is considerably smaller than the $5.4 billion royalty relief and drilling incentive package put in place by the Alberta government on 13 April 1982, under its Oil and Gas Activity Program (OGAP). Most of the credit, therefore, for any improvement in netbacks that the federal government might wish to attribute to itself will surely come from the Alberta government's own package, which was hardly given any recognition at all in the 1982 *NEP Update*.

The spring 1983 reduction in the world market price of crude oil necessitated a re-opening of the fall 1981 energy agreement, in large part because the wellhead price of conventional crude oil had breached the 75 per cent of world price cap. Although a roll-back in this price was threatened, the fact that government revenues from both oil and gas would thereby have been reduced made it unlikely. In any event, a Canada/Alberta Amending Agreement covering (at least) the period 1 July 1983 to 31 December 1984 was signed by Mr Chretien and Mr Zaozirny on 30 June 1983.

The main thrusts of this agreement were the following:

(1) to freeze the wellhead price of conventional crude oil (COOP) discovered before 1 April 1974 at $29.75 per barrel while this price lies within the 75–100 per cent of world price band;

(2) to freeze the Toronto city-gate price of natural gas at 65 per cent of the blended price of oil, with the NGGLT being finally reduced to zero on 1 February 1984, to accommodate a smaller-than-scheduled increase (i.e. more or less 16c per mcf rather than 25c per mcf) in the Alberta border price of natural gas, which will then remain unchanged until at least 1 February 1985;

(3) to extend the NORP to all oil discovered after 31 March 1974 (SOOP Oil), and to all production from infill drilling within pre-NORP entities;

(4) to maintain a substantial PCC and COC in effect until further notice, thereby continuing the sizeable wedge between the conventional wellhead price and the blended price of oil; and

(5) to leave intact the existing provincial royalty rates and federal PGRT rates for the foreseeable future.

The fact that this agreement took place with a minimum of public confrontation, and the fact that it represents to all concerned a reasonably sensible compromise on the difficult issues of oil and gas pricing and taxation, should reduce the uncertainty and instability of industry expectations that has without doubt characterised the last three years. The main problems remaining are the pricing and marketing of natural gas for export to the various regions of the United States, and the current level of the PGRT, which will become increasingly onerous if (as projected in Table 7.3) real netbacks decline from 1983 onwards.

Table 7.3 does, however, confirm that real (and nominal) dollar netbacks for both oil and natural gas were seriously eroded in 1981 from 1978–80 levels by the impact of the NEP, and especially the PGRT, on the producing industry. Indeed, relative to 1980, and with the possible exception of 'new-old' oil, these netbacks on an effective tax basis were eroded in real terms by more than one-third on EMR's own estimates. Even with all the new measures put in place, it is now projected

Table 7.3: Netback Calculations for Conventional Crude Oil and Natural Gas Produced in Alberta (constant 1981 dollars)

Year	Old-Old Oil ($/barrel)	New-Old Oil ($/barrel)	NORP Oil ($/barrel)	Old Gas ($/mcf)	New Gas ($/mcf)
			Large Crown Producers		
1975	5.23	6.52		0.41	0.46
1976	5.09	6.57		0.59	0.72
1977	5.33	7.07		0.71	0.88
1978	6.07	8.08		0.80	1.01
1979	6.13	8.04		0.90	1.15
1980	6.49	8.53		1.05	1.36
1981	4.20	7.20	6.80	0.61	0.90
1982	5.88	9.35	13.49	0.70	0.92
1983	7.22	11.18	11.39	0.75	0.93
1984	6.33	10.82	9.82	0.71	0.89
1985	5.75	10.36	9.39	0.71	0.90
1986	5.00	10.93	9.45	0.74	0.94
			Small Crown Producers		
1983	9.79	14.14	14.61	1.01	1.20
1984	8.80	14.00	12.77	0.98	1.16
1985	8.02	13.43	12.25	0.98	1.17
1986	7.08	14.01	12.36	1.02	1.22

Notes:

1. Source: These figures were obtained from officials at the Department of Energy, Mines and Resources, Canada. These netbacks are calculated on an 'effective' tax basis and imply certain reinvestment assumptions that may not materialise, and, indeed, could be quite misleading if interest rates are high and volatile. Netbacks on a full-tax basis (where corporate profits taxation rates applied are considerably higher) are much lower than these figures throughout. The lower portion of the table indicates that small producers receive larger benefits from the PGRT tax credit contained in the 1982 *NEP Update*. The Consumer Price Index has been used as a deflator throughout.

2. Old-old (COOP) oil refers to oil discovered after 31 March 1974. New-old (SOOP) oil refers to oil discovered after 31 March 1974 but before 1 January 1981. The same 31 March 1974 break-point divides old natural gas from new natural gas. NORP oil refers to all oil discovered after 1 January 1981, and to certain categories of tertiary and synthetic production begun before that date. Under the terms of the Alberta-Ottawa Amending Agreement of 30 June 1983, the new oil reference price (NORP) was extended to all oil discovered after 31 March 1974 (new-old oil), and to production from infill drilling within pre-NORP entities. Alberta royalty rates differ on these five categories of primary energy, and it is notable that Alberta's Oil and Gas Activity Program (OGAP) of April 1982 allocated largest royalty reductions to old-old oil and to old natural gas, for which the NEP had the most severe netback eroding effects (35 per cent and 42 per cent, respectively) in real terms. Average Alberta royalty rates for these five categories of primary energy after OGAP are as follows:

Old (Pre-1974) Oil 37%
New (1974–80) Oil 25%
NORP (Post-1980) Oil 23%

Old (Pre-1974) Gas 41% (38% if low productivity well)
New (Post-1974) Gas 33% (30% if low productivity well)

3. These netback calculations are based on the Amended Canada/Alberta Memorandum of Agreement of 20 June 1983, and therefore include the reclassification of SOOP to NORP as well as the important effects of Alberta's OGAP programme. They therefore differ from the expected netbacks that may have been perceived by the industry from the vantage point of 1981 or 1982. As a check on these EMR netback figures, comparisons were made with those compiled in the latest (August 1983) Lewis Engineering Profitability Analysis Service report. Although there were numerous discrepancies, overall both series proved to be consistent with each other.

that netbacks on 'old-old' oil will not be restored to 1980 levels in real terms until the latter half of 1983. On old and new gas, real dollar netbacks are not restored until after the end of the energy agreement, if at all, unless there is a totally unexpected rebound in the marketability of Canadian gas in the US, which would affect netbacks through the revenue 'flow-back' system currently in place.

Unless one believed that netbacks were much too large in the 1978–80 period, and that the erosion of them would not affect producers' expectations and confidence, the consequences of the severe real netback erosion for exploration and development activity at a time of high real interest rates should have been at least broadly anticipated. Cash flows from existing production are important for firms to extend their exploration and development activity, since these cash flows largely determine their ability to borrow on either debt or equity account.

The recent extension of NORP oil prices to SOOP oil and to oil produced from newly-drilled infill wells will provide greater cash flow to producers. Nevertheless, in so far as netbacks have gradually been restored, it has largely been through provincial royalty relief. Implicitly, therefore, the Alberta government is now paying a significant proportion of the PGRT out of its own revenues.

As we shall argue in the next section, just as it bought back jurisdictional control over its own resources by agreeing to pay the Petroleum Incentive Payments (PIPs) on provincial lands in exchange for inducing the federal government to apply a zero rate of natural gas tax on exported gas, the Alberta government has now attempted to buy back some moderate prosperity for the industry by providing royalty relief to help bring real netbacks back towards their pre-PGRT standing. The consequences of all this for revenue distribution are not inconsiderable, and will be discussed in more detail in a later section.

Jurisdictional and Related Issues

In the post-NEP, pre-NEA interlude, it became clear to most observers that prices for petroleum and natural gas and the division of revenues were not the sole concerns of the two levels of government, particularly the provincial governments. The latter perceived the NEP as an undisguised attempt by the federal authorities to enhance their jurisdiction over resource development in this country (at least oil and gas resources), a contravention of provincial rights under the British North America Act (and now the Constitution of Canada). The decision by the federal government to tax not only domestically used natural gas, but also gas exports, was particularly offensive to the provinces.

Alberta responded to the NEP by delaying approval of new oil sands plants and reducing oil production, thus asserting provincial control over the pace of resource development and exploitation in the province. Since this response meant that greater imports of crude oil were necessary, it was also a way of putting pressure on the federal government to relax its stance.

Under the terms of the NEA, Ottawa withdrew the natural gas and gas liquids tax on exports of natural gas (while still asserting the right to levy it). This provision, along with the higher wellhead gas and oil prices, was a key factor in permitting an agreement to be reached. Since about one-third of all gas produced is currently exported, the removal of this tax improves the returns to gas producers and reduces potential federal revenues.

Concurrently, the government of Alberta took over the administration and financing of the federally-based PIPs for exploration work within the province, thereby seeming to regain more complete control over provincial energy development. This greater involvement is expected to cost the province $2.8 billion over the life of the agreement, which seemingly offsets the reduction in federal tax revenues from natural gas exports. In retrospect, however, this may have been a poor financial deal from the perspective of the province. Indeed, if the tax on natural gas exports had been kept level with the tax on domestic gas, it would not now be projected to earn much revenue in any case, indicating that Alberta paid a heavy price for retaining jurisdictional control over the PIPs.

This reallocation of responsibilities for PIPs still does not alter the fact that the incentives for exploration and development on federal lands, principally in the Arctic and offshore on the east coast, will be higher than those in the conventional oil- and gas-producing areas. In effect, higher cost frontier resource developments are being encouraged at the expense of potentially lower cost ones in the western Canadian sedimentary basin, a situation which makes little sense from the point of view of the optimal timing of exploration and development of potential oil and gas pools. From this perspective, one might interpret the $5.4 billion of Alberta incentives, including both drilling incentives and royalty relief, partly as an attempt to redress this imbalance while also shoring up industry revenues more generally. Indeed, this may be exactly what Ottawa had hoped would occur.

Two comments remain to be made on the PIPs question. First, to the extent that Alberta-financed incentive payments are capitalised into land lease and bonus bids, some of these payments will return to the provincial government. In any event, whether the incentives turn out to be too generous or too lean, the consequences are internalised to one level of government through these capitalisation effects. Secondly, in taking over the PIPs programme, the Alberta government has been involved in an exercise in buying back jurisdiction control over its own resources. Saskatchewan and British Columbia, which are letting Ottawa administer and finance the PIPs within their own borders, have apparently not felt the issue was as important to them.

Some observers have taken the view that the Ottawa-Alberta jurisdictional dispute is to be blamed for the cancellation of construction of additional oil sands plants, most notably the Alsands plant, and that if approval had been given sooner, construction would have commenced and would be continuing now in spite of high interest rates and the slowdown in the advance of world oil prices. It is true that soon after the unilateral publication of the NEP in October 1980, Alberta announced that approval for new oil sands plants would be delayed until a comprehensive energy arrangement was negotiated with the federal authorities. Yet, when one examines the April 1982 generosity of the governmental assistance offered to Alsands, the last proposal to remain potentially active, this view is not easily upheld.

The two levels of government were willing to provide up to 50 per cent of the share capital, and each make loan guarantees of 34 per cent of the private participants' pre-production outlays. No payments would be required until after start-up, while at the same time no tax or royalty payments would be due until the loans were repaid. These and other provisions would have provided the private oil companies at least a 20 per cent nominal return on their investment. It appears likely that if these provisions were not sufficient inducements to cause the partners to proceed, they would not have been enough to keep them going even if they had commenced earlier. The uncertainty regarding future world oil prices, possible concerns about whether the two governments would change the rules of the game in midstream, coupled with exceptionally high interest rates, have been the main deterrents, not merely the federal-provincial jurisdictional disputes *per se*.

Another dimension to the Ottawa-western Canada conflict arose two years ago. It had to do with the shut-in production capacity of oil in western Canada while extensive imports by the Montreal refineries were continuing. Although the now rescinded Alberta production cutbacks may have stimulated this situation, it should have quickly rectified itself.

In the 1982 *NEP Update*, the federal government argued that there was no overall incentive for eastern refineries to buy foreign crude oil instead of western oil because, if imports were lower priced, it would lower the mean cost per barrel of imports and hence reduce the average federal subsidy paid to refiners of foreign crude. It admitted, however, that if individual refiners can get foreign shipments at below the average price, they can benefit significantly from so doing; and it allowed that before 1 April 1982 compensation per barrel was set prior to the month when the oil was imported. The implication of this was that if *all* refiners could buy spot shipments at below the projected price for compensation purposes, *all* of them could gain from this lag in the compensation formula.

To the extent that shut-in capacity in western Canada is allowed to continue, all Canadians lose as foreign producers receive revenues that could have been distributed among the domestic industry, provincial and federal governments and Canadian consumers. Some measures, including removal of the lag in the compensation formula and increased exports of certain categories of crude oil, have been announced by Ottawa to minimise these losses.

An additional area of potential disagreement between Ottawa and Alberta concerns natural gas exports to the United States. These have recently been running much below both contracted levels and the potential of the industry. In its statement on 13 April 1982, Alberta suggested a number of changes to enhance gas exports. First, the National Energy Board's 'short-term deliverability test' was labelled as too restrictive and inconsistent with the historical record, since it assumed that no increments to reserves of natural gas would occur during the next five years. The recommendation that this assumption be relaxed has now been implemented.

Second, because in many instances the quantities of gas being exported are substantially less than those contracted for, a realistic assessment is being made of the amounts that will actually be taken under existing contracts so that the balance can be freed for delivery under other contracts. Flexibility in price and contract conditions, depending upon where in the USA the sales are to be made, also seems to be required in order to compete with existing local market situations. The current one-price policy on all gas experts does not allow for this leeway, although an 11 per cent reduction in that price has already been implemented. Whether or not a further reduction would expand natural gas export revenues remains a moot point, though it is clear that such an expansion would be an additional method of stimulating further exploratory and development activity by the oil and gas industry. Under pressure from the US authorities, the export price has been lowered a further 25 per cent on all quantities exceeding 50 per cent of existing contract volumes. The effects of this reduction, as yet, remain unclear.

Finally, there remain jurisdictional problems with respect to offshore resources. Although a Canada-Nova Scotia Agreement on Offshore Resource Management and Revenue Sharing was signed on 2 March 1982, satisfactory arrangements between the federal government and Newfoundland for the development of the Hibernia oil field are, despite a Supreme Court ruling in Ottawa's favour, awaiting further inter-governmental negotiations.

A negotiated solution is badly needed to the jurisdictional and revenue-sharing dispute between Ottawa and Newfoundland over the Hibernia and other offshore oil fields. Given the fact that local jurisdictional control is bound to be necessary to some extent, and the fact that Newfoundland is currently Canada's least prosperous province, I am one of those who would err on the side of generosity towards the Newfoundland position. Given a relatively flat profile for world oil prices, and the high cost of exploration and development in the Northern Atlantic waters, the expected economic rents obtainable from Hibernia and other offshore developments cannot be so substantial that it would make that much difference to the average Canadian taxpayer whether the federal government captured a share of these rents or not. In other words, Newfoundland should be permitted to harness this resource development opportunity to its own abundantly evident fiscal, income-generation and employment needs.

Revenue Sharing

The energy agreements signed in the fall of 1981 between each of the three western producing provinces and the federal government were expected to yield substantial additional revenues to the two levels of government as well as to industry, compared to those projected in the NEP. With the subsequent levelling out and reduction in world oil prices, however, the updated total returns to the three parties are expected to be down by some 37 per cent compared to the 1981 agreements. The largest drop will be experienced by the federal government, with the provinces (chiefly Alberta, which generates between 85 per cent and 90 per cent of all oil and gas revenues) not far behind. Industry revenues after operating costs will be down by a smaller percentage (see Table 7.4).

The reasons why the governments bear the largest portion of the expected decline are twofold. Revenue from their taxation arrangements, which are essentially designed to skim off the economic rent or return over and above that necessary for the industry to continue supplying oil and gas to the market-place, are naturally reduced as anticipated prices and therefore economic rents diminish. Secondly, 1982 saw reductions in tax and royalty rates and improvements in incentive systems by the Alberta and federal governments estimated at $5.4 billion and $2.0 billion, respectively. These were conceived to soften the adverse effects for the industry of the 1981 federal-provincial energy agreements and the subsequent reduction in the anticipated rate of oil price increase.

The NEP and the 1981 agreements reduced the federal dependence upon the corporate income tax for extracting revenues from the petroleum and natural gas industry. Yet the recent tax changes have restored somewhat the prominence of the corporate tax as a federal revenue source. The increase in the estimated share of net Ottawa revenues coming from the corporation income tax may well be due to a recognition by federal authorities that this tax is a more efficient revenue raiser than the IORT.

The figures in Table 7.4 are somewhat misleading, however. On the one hand, they may be overly optimistic with respect to the revenues that the provinces, especially Alberta, can expect to receive from land sales. At the same time, they make no allowance for required infrastructure investment costs borne by provincial governments to support industry activity. On the other hand, the federal revenues are understated (and industry revenues overstated) since the federal PIPs on Canada Lands have no explicit place (and do not belong) in a revenue-sharing table pertaining to revenue shares from production on provincial lands.

In addition to this, the consumer share of *potential* net revenue is omitted from the table. Notice, first, that the federal share nets out the subsidy on both imported and new oil from the petroleum compensation charge (PCC) receipts. The PCC receipts are excluded on the grounds that they are not intended to exceed the amount needed to subsidise consumers using imported oil or new oil eligible to receive the world price. It is appropriate to exclude that portion of the PCC used to provide the world price to new domestic production, including synthetic production. Such payments have already been included in the tabulations showing provincial royalties, industry receipts and the federal income tax. But the share designated to reduce the price of imports is essentially a tax the federal government is levying to consumers of imported petroleum products. Over the five years of the agreement, the subsidy on imported

Table 7.4: Revenue-sharing Estimates (billions of dollars)

	Sept. 1981 to Dec. 1986 Totals				1981 to 1984 Annual Figures			
	Energy Agreements Autumn 1981	Energy Update Spring 1982	Revised Estimates Fall 1983	Per Cent Change From 1981	1981 Annual	1982 Annual	1983 Annual	1984 Annual
	(1)	(2)	(3)	(4)	(5)	(6)	(7)	(8)
Government of Canada								
Canadianisation Levy	1.5	1.6	4.7		0.6	0.9	0.7	0.6
Natural Gas and Gas Liquids Tax	13.8	4.1	2.2		0.8	1.2	0.6	0.0
Oil Export Tax	0.7	1.6	1.3		0.4	0.3	0.1	0.0
Incremental Oil Revenue Tax	7.1	3.7	0.5		—	0.3	0.1	0.3
Petroleum and Natural Gas Revenue Tax	23.0	16.6	10.9		1.1	1.7	1.9	2.1
Petroleum Incentive Payments	(7.4)	(8.9)	(7.7)		(0.5)	(1.1)	(1.5)	(1.5)
Corporate Income Tax	22.0	16.9	15.9		1.3	1.6	2.2	2.3
Surplus Petroleum Compensation Charge	0.1	0.0	0.0		(0.1)	0.3	(0.1)	(0.1)
Land Payments	0.0	0.0	0.0		0.0	0.0	0.0	0.0
Sub-total	60.8	35.6	27.8	54.3	3.5	5.1	4.0	3.7
Per Cent of Total	(28.5)	(22.0)	(20.6)		(23.0)	(27.4)	(20.0)	(18.4)

Provincial Governments								
Royalties and Freehold Tax	64.0	42.2	30.0		4.6	5.3	5.8	6.0
Land Payments	9.7	9.7	2.9		0.8	0.5	0.7	0.7
Oil Export Tax	0.7	1.6	1.3		0.4	0.3	0.1	0.0
Corporate Income Tax	4.3	3.3	3.1		0.2	0.4	0.4	0.5
Petroleum Incentive Payments	(4.3)	(4.3)	(2.8)		(0.4)	(0.5)	(0.6)	(0.8)
Royalty Adjustments	0.0	0.0	0.0		0.0	0.0	(1.0)	(1.0)
Sub-total	74.4	52.5	34.5		5.6	6.0	5.3	5.4
Per Cent of Total	(34.8)	(32.4)	(25.6)	53.6	(37.3)	(32.3)	(26.7)	(26.6)
Industry								
Cash Flow	76.4	70.3	64.8		5.9	6.4	8.1	8.4
Petroleum Incentive Payments	11.6	13.2	10.5		0.9	1.6	2.1	2.3
Land Payments	(9.7)	(9.7)	(2.9)		(0.8)	(0.5)	(0.7)	(0.7)
Other Adjustments	0.0	0.0	0.0		0.0	0.0	1.0	1.0
Sub-total	78.3	73.8	72.4		6.0	7.5	10.6	11.1
Per Cent of Total	(36.7)	(45.6)	(53.8)	7.5	(39.7)	(40.3)	(53.3)	(55.0)
Total Revenues	213.6	161.9	134.7	36.9	15.0	18.7	19.9	20.2
	(100)	(100)	(100)		(100)	(100)	(100)	(100)

Table 7.4 continued

Note:
Columns 1 through 6 are based upon revenue sharing for the entire Canadian petroleum industry, whereas Columns 7 and 8 refer only to the Alberta sector, which accounts for approximately 85 per cent of total Canadian production. In our view, the industry share from Alberta production in columns (7) and (8) is overstated, since it should only include Alberta's PIPs and not the federal PIPs payable on exploration and development activity in other provinces and (more especially) Canada Lands. Similar reasoning suggests that the federal share is equivalently understated.

Sources:

Columns 1 and 2: Canada, Department of Energy, Mines and Resources. *The National Energy Program: Update*, 1982. Information Notes, Table 1.

Column 3: Canada, Department of Finance, *The Fiscal Plan*, background paper to the Budget, 19 April 1983, Tables 2.5 and 3.1. Alberta Treasury, 1983 Budget Address, 24 March 1983, Table A2 and author's own estimates.

Columns 5 and 6: Petroleum Monitoring Agency Canada, *Canadian Petroleum Industry, Monitoring Survey*, 1982. Ottawa: Supply and Services, Canada, Table 8, p. 5–2.

Columns 7 and 8: Figures were provided by officials at the Department of Energy, Mines and Resources, Canada.

oil is expected to average around $1 billion per year.

In addition to this, notice that the Ottawa policy of keeping prices of *domestically produced* old oil and natural gas below world market prices is also equivalent to taxing producers of oil and gas (and especially the producing provinces) and using the tax to subsidise consumers. Therefore, it makes sense to count the value of this 'tax' as part of the consumer revenue share as well. Even if it is assumed that the market price for natural gas at the US border is at best only 75 per cent of the BTU equivalent world price of oil, the value of this combined federal 'tax' on oil and gas is substantial over the life of the New Energy Agreement, averaging about $3 billion per year. The value of this benefit to consumers of subsidised oil and gas divides about equally between the oil and gas accounts. The largest portion of it has, however, already accrued during the 1981–3 period.

By permitting the federal government to improve markedly its revenue share in the energy agreement, and by making its 1982 reductions in net taxes on the industry approximately three times the amount the federal government has granted, the Alberta government has, at least implicitly, agreed to make considerable contributions from its potential non-renewable resource revenues to the federal government and Canadian energy consumers. These contributions represent a clear-cut bottom line for

(a) getting the economy moving again;
(b) recognising the revenue distribution problem; and
(c) re-establishing jurisdictional control over provincially-owned resources.

Thus, a rent-sharing scheme has now implicitly been put in place, in which Alberta has already made major concessions in sharing potential petroleum and natural gas revenues with the rest of Canada. No one should now expect Alberta to contribute generously to a further inter-provincial rent-sharing scheme, as some have suggested should be the result of continuing negotiations on federal-provincial fiscal arrangements. Other provincial jurisdictions with substantial hydro-electricity generating capacity capture large-scale economic rents for the public sector, and/or choose to distribute them to consumers within the province instead. These must also be accounted for in any revised federal-provincial equalisation scheme.

With the recent decline in oil industry activity and in the buoyancy of its oil and gas revenues, the government of Alberta now finds itself with a budgetary situation much more like that in other provinces. Indeed, it has been running an annual General Revenue Fund (GRF) deficit for two years now, and will continue to do so in the current budget year. This deficit is being covered in part by taking all the nominal interest earnings (projected at about $1.5 billion) out of the Alberta Heritage Savings Trust Fund (AHSTF), in part by cutting the non-renewable resource revenue transfer to the AHSTF from 30 per cent to 15 per cent of net revenues received (about $650 million), and in part by borrowing from the money market. In fact, the remaining resource revenue transfer of about $650 million will hardly be sufficient to keep the nominal AHSTF capital sum of $13.7 billion (or 1.4 times current annual GRF expenditures) from falling in real terms. Thus, as recent events have demonstrated, Alberta has now reached the point at which provincial tax rates, especially including the provincial personal income tax rate, must begin to be increased if the government is to maintain its pattern of outlays on the provision of public goods and services to Alberta residents.

Put differently, the Alberta government accounts are now in significant real (or inflation-adjusted) deficit even when the AHSTF is taken into account along with the GRF. A considerable expansion in natural gas exports might put them back into substantial surplus in the mid-1980s, but for the foreseeable future it is no longer possible to argue that Alberta's fiscal capacity is much larger than that of other Canadian governments. Moreover, if the recent collapse in Alberta's overall rate of economic growth, and virtual tripling of her unemployment rate despite a cessation in net immigration, are any indication for the future (and the cancellation of the Alsands and Cold Lake projects, the postponement of several large-scale petro-chemical developments and, especially, the projection of much flatter world energy prices would suggest that they are), we should no longer expect the Alberta economy, or the western region economy more generally, to out-perform the rest of the Canadian economy by anything like the wide margins that it did prior to the fall in 1980; indeed, quite the opposite seems likely to occur. That will in part be due to the aftermath of Ottawa's National Energy Program. But it will also be due to the fact that, like all the OPEC countries, our main petroleum-producing region will fall short on relatively less buoyant times as the world supply-demand balance for fossil fuels makes flatter real energy prices and netbacks the rule for the next few years. Indeed, the Alberta economy will remain sluggish until such time as several additional tar sands and heavy oil developments of medium size become economic to build, or until US gas markets pick up, perhaps in mid-1986.

Foreign Ownership, Canadianisation and the National Energy Program

One of the objectives of the NEP has been the Canadianisation of the petroleum and natural gas industry. The target is for 50 per cent of oil and gas production to be Canadian-owned by 1990. The elimination of tax write-offs and the replacement of them with drilling and other exploration incentives (PIPs), which are substantially greater for companies having a high percentage of Canadian ownership, is a major policy change designed to assist in achieving this objective. In consequence, foreign-owned firms now find it more difficult to continue their exploration incentives. At the same time, their current and anticipated netbacks have been eroded through the new pricing and taxation regime. The resulting impact on their share prices has lowered the costs of buying them out for Canadians. Alberta has explicitly accepted the Canadianisation objective by agreeing to administer and finance the PIP programme within the province.

The Canadianisation programme has had some success. Between the time the NEP was first announced in October 1980 and early 1983, Canadian ownership of the industry had increased from 28 per cent to nearly 35 per cent. Over a dozen sizeable foreign corporations were acquired by Canadian interests during this time period, and two domestic corporations, Petro-Canada and Dome, are now among the top ten companies in terms of upstream revenues. However, much of this shift in ownership reflects trends which began well before the NEP was announced.

This change has not been without its problems. In particular, the United States has perceived the incentives programme to imply outright discrimination against foreign companies, and has proposed that such discrimination be removed. It has also objected to the 25 per cent back-in provision for Petro-Canada on potential frontier developments on Canada Lands. Concurrently, it has raised concerns about the Foreign Investment Review Agency (FIRA) with its requirement that foreign takeovers of Canadian firms, or even of foreign subsidiaries in Canada, clearly show significant benefits to Canada. So far, however, any US retaliation to these Canadianisation objectives has been too subtle and diffuse to identify clearly.

A second issue concerns the possible downward pressure on the Canadian dollar that results from the outflow of funds involved in the purchase of domestic subsidiaries of foreign oil companies. The magnitude of such purchases has been perhaps as high as $15 billion since the NEP was announced. This appears to have been part of the reason for the Canadian dollar's weakness during 1981. A recent study by the Bank of Montreal (1981) suggests that the dollar may have been weakened by 1½ US cents between October 1980 and June 1981, as a consequence of take-over activity. This estimate is based on the assumption that even if the companies buying the foreign subsidiaries were to borrow US dollars from Canadian chartered banks (which at the end of July 1981 had over 40 per cent of their total assets, or $140 billion, in foreign currency) and use these to pay the foreign shareholders, the banks would have wanted to cover themselves in the forward market by selling Canadian dollars. In turn, the downward pressure that would be exerted on the Canadian currency in the forward market would be felt in the spot market. The study also suggests that, because exchange market participants would expect there to be some such pressure on the dollar, such anticipations would, in themselves, work towards their own realisation.

In the future, the debts must be serviced; to the extent that interest payments are greater than the dividend payments foreign-owned firms would previously have made to their shareholders abroad, downward pressure on the Canadian dollar will ensue. The Bank of Montreal study estimates that if the pace of takeover activity that occurred up to June 1981 (about $2 billion per quarter) continued for the entire year, the downward pressure on the Canadian dollar would peak in about two years at 1¼ US cents. If the takeover activity continued for three years at about $1.6 billion per quarter, the Bank's model suggests that the Canadian dollar would settle about 5 US cents below what it would otherwise have been. However, it is not clear that this work adequately reflects the concurrent changes taking place in the Canadian economy, such as the reduction in oil imports that should result from higher domestic oil prices. These would place partially offsetting upward pressure on the Canadian dollar.

Nevertheless, it is unfortunate that the take-over activity stimulated by the discriminatory tax and incentives system instituted under the NEP came at a time when it was difficult enough to manage our monetary and exchange rate policy in the face of record high US interest rates. The resulting softness in the Canadian dollar added, perhaps unnecessarily, to both our overall inflation rate and our real interest rates. Indeed, the Canadian dollar did fall by 4½ US cents over the two years subsequent to the NEP, and the uncovered interest rate differential widened from between 2 (for long-term yields) and 3 (for short-term yields) percentage points above its historic norm during a similar period.

The point needs to be made, however, that if less dependence on foreign capital is desired in the longer-term, greater savings by governments and the private sector are essential. It makes little sense to Canadianise the oil and gas industry at the expense of de-Canadianising other industries in the process. Artificially stimulated take-overs of existing foreign-owned assets neither create new producing assets, nor are they necessary to stem some hypothetical outflow of potential oil and gas rents when alternative fiscal measures are at hand if required.

Macroeconomic Issues and Energy Developments in the 1980s

Once an energy accord had been reached between Ottawa and the oil- and gas-producing provinces, the energy sector might have expected some assurance that many of the important rules of the game would not be changed for a five year period. The major item of doubt remaining in the accord was the course of inter-national prices for oil. Since these did not increase as hypothesised, it was to be expected that some important re-negotiations of taxes and royalties would be bound to occur, particularly if at least some large-scale energy projects were going to proceed.

The macroeconomic implications of large-scale energy-related investment projects are considerable, but it is outside the scope of this paper to attempt any thorough discussion of them. Suffice it to say that care will have to be taken to ensure that these expensive projects are not subsidised so heavily that they proceed while lower priced conventional oil and gas development is neglected; that they are scheduled so as not to create serious sectoral bottlenecks and additional severe inflationary problems; that financing of them is accomplished in a way that both encourages and utilises domestic savings and minimises foreign exchange and balance of payments disruption; that appropriate purchasing policies are employed which reach a reasonable balance between importing cheaper foreign equipment and stimulating technological development and employment domestically; and that the potentially enormous pollution effects of such plants are controlled. Important inter-regional questions also arise with respect to the economic linkages or inter-regional spillovers from potential large-scale energy projects, including those regionally-specific impacts which work through the Canadian balance of payments and the international value of the Canadian dollar.

Important questions can be raised concerning the collection and distribution of economic rents from energy resource developments (and this pertains as much to Quebec's hydro-electricity potential as it does to western Canada's fossil fuel potential). Indeed, a serious study needs to be made to compare the implications of various rent-distribution schemes for intra-regional and inter-regional income distribution and the efficiency of labour and capital market responses. The provincial under-pricing of electricity, natural gas or petro-chemical feedstocks has different developmental and efficiency implications than the lowering of provincial labour income taxes and/or sales taxes, implying the under-pricing of government services more generally. There may also exist an optimal trade-off between the resource allocative inefficiencies resulting from preventing the inter-regional terms-of-trade from adjusting as far as they ought, and those which may result from fiscally induced migration.

It has not been possible in recent years to separate energy policy developments fron overall fiscal policies. Moreover, since energy price changes do impact on the overall rate of inflation, and on the level of unemployment in both energy-producing and energy-using sectors and regions, monetary and exchange rate policies do not remain immune from energy questions either. For example, whenever there is a substantial change in the inter-regional terms of trade, the monetary and exchange rate policy which may seem to be most appropriate from the perspective of an energy-using region may well look to be somewhat inappropriate from the perspective of an energy-producing region, or vice versa. In any case, the macroeconomic implications of energy pricing policies will inevitably depend upon what concurrent monetary policies are implemented.

As far as fiscal policy is concerned, it is crucial to ensure that the fiscal regimes faced by all investment projects are reasonably comparable, so that lower cost resource developments proceed before higher cost ones. To provide higher prices or lower taxation and royalty arrangements for more risky projects (let us say because they are on Canada Lands rather than provincial lands) will only distort the time profile of resource development completions, leading to serious economic inefficiencies. It is for this reason that corporate profits taxation, when combined with the absence of special tax concessions and write-offs, is to be preferred to revenue taxation of the form exemplified by the PGRT and IORT.

The less interventionist we can become, and the smaller the number of specially designed tax levers and incentive schemes pertaining to particular industries, the better will the goal of economic efficiency be served. Indeed, it is now time to consider dismantling all the heavily interventionist mechanisms contained in the NEP.

The Economic Council of Canada and the C.D. Howe Research Institute are just two of the influential agencies which have recently recommended that the domestic wellhead price for conventional crude oil now be raised to world levels in light of the current flatness in world oil prices. This would help to ease the remaining damage done to the energy sector by the massive sectoral tax increase imposed in 1981 through the impact of the NEP. Indeed, the rationale for this tax increase can only be explained if it is clearly understood that the main objective of the NEP was to transfer a substantial proportion of resource revenue potential out of the province of Alberta. Since the justification for this wealth transfer is fast disappearing, and since the NEP has also had demonstrably perverse effects on both economic efficiency and security of energy supply, it is now time for Ottawa to deregulate the energy sector. In my view, the numerous changes that have had to be made to the NEP since its inception clearly demonstrate how misconceived it was.

Put differently, it is impossible to avoid the conclusion that the fundamental purpose of the NEP was really to create and sustain a large-scale transfer of wealth from the Province of Alberta to the rest of Canada. Judged solely by this principal aim of wealth redistribution, the NEP has been successful. Since significant wealth destruction has occurred in the process, however, this redistributive objective had definitely been a negative-sum game.

Moreover, if world oil prices continue to be flat in nominal US dollar terms, and if natural gas markets in the United States do not rebound substantially, then there will soon no longer be much room in the fiscal system for any special federal taxation arrangements pertaining only to the primary oil and gas industry. These levies include the petroleum and natural gas revenue tax (PGRT), the incremental oil revenue tax (IORT), the natural gas and gas liquids tax (NGGLT) and the Canadian ownership charge (COC). On the other hand, the federal government should *not* be financing special industry incentives, including both depletion allowances and petroleum incentive payments (PIPs), to the extent that it is now doing on both provincial and federal lands. (It is not paying for PIPs in the Alberta segment now in any case.) On balance, with provincial royalties being non-deductible, this would still leave the federal government with a significant positive share of revenues from the industry through the regular channels of the corporate income tax.

Such an elimination of substantial federal distortions, especially if coupled with price deregulation by the federal government, would go a long way to improving overall industry efficiency and the prospects for security of energy supplies. In other words, since the wealth redistribution argument in favour of the NEP is fast disappearing, my bottom line conclusion is that after three years of costly experimentation the NEP should now, in large part, be dismantled.

DISCUSSION

The speakers' comments stimulated considerable discussion and debate. A number of questions were posed from the floor, and answered by the speakers in the discussion summarised below.

(1) It was pointed out that political and economic factors could never be treated in isolation from each other. Thus, Canadian energy policy could not be divorced from Canadian Middle East policy in general, and Canada's stance on the Iran-Iraq war and on the question of Palestine in particular. This close relationship follows from the fact that Canadian energy policy is premised on assumptions about world oil supply and world prices. Speakers from the floor called for Canada to take a more even handed attitude towards issues in the Middle East, and a more balanced role that would be compatible with its own long-term national interests.

(2) Deputy Minister Tellier responded by insisting that the government of Canada does not dictate to private sector interests the sources from which they should purchase oil. Only in state-to-state deals, he insisted, was there any scope for tailoring commercial to political priorities. And Canada engages in such state-to-state deals only with Pemex, the Mexican state oil comapny.

However, some members of the audience reacted sceptically to his comments.

(3) In a similar way, it was noted that since the oil price adjustments of 1973–4 there have been no such things as regional solutions to energy supply problems. Global problems require global solutions; and Canadian policy-makers can no longer assure their energy supplies by sticking their heads in the sand whenever the world political situation looks ominous.

(4) It was pointed out that export trade between Canada and the Gulf has increased considerably in recent years. The Arab countries, attracted by Canadian technology, resources, and lack of the taint of imperialism, have been deliberately attempting to increase their co-operative economic relations with Canada. But everything Canada can offer them is obtainable elsewhere. If Canada wants to maintain and increase the existing level of economic exchanges with the Arab countries, it will have to make its foreign policy more even-handed.

(5) Speakers from the floor pointed out that while the Arab Gulf states take a global view of energy security, Canada takes a very parochial one. The Arab states set oil policy in accordance with both their own national development requirements and the need for prosperity and stability in the world economy. It was the Gulf group of countries inside OPEC that assured stable prices, and therefore assisted Canada and other non-OPEC producers in maintaining their existing level of production. Furthermore, Canadian oil self-sufficiency exists only in the aggregate. It still must import from the Middle East to offset the western Canadian oil sent to the US. Despite Canada's nominal self-sufficiency it must be always conscious of the political situation inside the Gulf region and of how its foreign policy stance in the Middle East is perceived and received.

BANQUET REMARKS

Dr Alvin Lee,
President and Vice-Chancellor,
McMaster University, Canada

Your Excellency Sheikh Nasser Mohammed Al-Ahmed Al-Sabah
Regional Chairman, Mrs Jones
Dr Kubursi
Other Distinguished Head Table Guests
Members of the Petroleum Information Committee
Ladies and Gentlemen

It is a great pleasure for me as President and Vice-Chancellor of McMaster University to welcome each and every one of you to this banquet and to this important symposium on the energy industry.

The symposium itself is a public service presentation of McMaster University and the Petroleum Information Committee. It is an honour for McMaster to be associated with this important committee. As some of you will know, the membership of the committee includes senior political and administrative officers from the Ministries of Energy and Information from the following countries: Saudi Arabia, Kuwait, Iraq, Bahrain, Oman, Qatar, and the United Arab Emirates. The commitee is based in Kuwait and was established in 1979. It is chaired by the very distinguished gentleman who is to address us tonight, Sheikh Nasser Mohammed Al-Ahmed Al-Sabah.

McMaster University's relationship with the Gulf States is not new. We are very proud that members of this university have worked in the area over a number of years. We have joined you in your efforts to bring about massive social and economic developments in your region of the world, and we wish you all success. My predecessor as President of

121

McMaster, Dr Arthur Bourns, was privileged to go a few years ago as one of two educational consultants from Canada to Oman to help plan the first university in that country, Kaboos University.

McMaster has a long-standing and very effective commitment to research. There are many instances in which members of the international community have called upon the excellence of our faculty. We are proud to share our knowledge and expertise, particularly with developing countries, and we hope that you will find many more instances to call upon us.

Although we have many students from the Arab world, we have only a few from the Gulf region. Given the interests and experience of many of our faculty — including some in economics, in engineering, and in health sciences — we are in a natural position to receive more students from your area. On your next visit to our university, we hope in the near future, we would like you to be able to see several of your students here. While our academic ambitions are large and we play in the big league in research and higher education, our hearts are warm and the McMaster campus is a friendly place.

It is fitting on this occasion that we recall an ancient connection between universities in the Western world and the Gulf Arab States. In the oldest building on the McMaster campus, called University Hall, there is a large, formal room known as Convocation Hall. Along its two outer walls are rows of stone heads representing the great philosophical and scientific thinkers of the Western world from Socrates, Plato, Aristotle, Bacon and Spinoza to Kant and from Hippocrates, Copernicus, Newton, Faraday and Darwin to Lord Kelvin. This intellectual history is one which connects closely at important points with Islamic traditions. The earliest of these thinkers, Socrates, Plato and Aristotle, Hippocrates and Copernicus are held in common by our two traditions. Not only does much of Muslim and Christian belief about education stem from them but the two traditions have, at different times through the centuries, vitalised each other.

We in the Western intellectual tradition, aware of the lines that reach from Athens and Jerusalem through classical, medieval, renaissance, and enlightenment Europe into modern universities like McMaster, need also to recognise that other great tradition of inquiry and enlightenment that had its first flowering in the Bayt Al Hikmah in Baghdad from AD 830 onwards. There, the philosophical and scientific texts of Greece, Rome, Persia and India were translated and absorbed. In the ninth and tenth centuries, colleges and centres of learning developed throughout the Muslim world, at Al Azhar in Cairo,

Qarawiyin at Fez in Morocco, Zaytouna in Tunis, and in Cordoba in Spain.

We need to recall that, in these centres, Muslim scholars excelled in astronomy, geography, physics, philosophy, mathematics and medicine. As a medievalist, I know well that much of what happened intellectually in the great European universities of the high Middle Ages — Bologna, Paris, Oxford — was fundamentally altered by the influence of Muslim scholarship. It was from that source that classical Greek thought re-entered the European tradition, understood, translated, and transmitted by men like Avicenna (at Seville and Cordoba) and Averroës. It is not an overstatement to say that the intellectual Renaissance in Europe would not have happened except for the Islamic infusion. Your Excellency, and our other guests from the Arab States, I wish *with you* to honour the historical intermingling of our respective cultural traditions and to celebrate with you that now again in the late twentieth century the associations are taking new and vital forms.

We Canadians are proud to belong to a country that has taken on major, active responsibilities for ensuring peace in the world. The Middle East countries have survived more than their fair share of crises and we sincerely hope that our governments can work together to bring peace to this vital and deserving region.

It is primarily oil and economics, important matters both, that have brought us together. But this symposium has also shown that human contact and dialogue are even more powerful forces in establishing good relationships and friendships between individuals and countries. McMaster University welcomes you most warmly and offers you ongoing friendship.

BANQUET SPEECH

His Excellency
Sheikh Nasser Mohammed
Al-Ahmed Al-Jaber Al-Sabah,
Minister of Information, Kuwait

Mr Vice Chancellor
Your Excellencies Arab Ambassadors
Mr Prudhomme
Mr Tellier
Your Worship
Distinguished Guests

Thank you, Dr Lee, for you gracious introduction. Permit me to take this opportunity of expressing our appreciation both to you and the organising committee, to the speakers and to all the participants in the symposium. You have made this gathering a most stimulating and rewarding experience for each and every one of us in the Petroleum Information Committee of the Arab Gulf States.

We are indeed pleased and satisfied to have co-operated with you in this fruitful effort and we look forward to further co-operation in the near future.

Ladies and Gentlemen: It is, of course, true that Canada and the Gulf are thousands of miles apart geographically. However, our concerns and interests are mutual and convergent. Canada's experience in harmonising the interests of oil-producers and oil-consumers is a model to follow on a world level.

If there is one fact that has been made abundantly clear by this symposium it is that the world does not have many years in which to make a smooth transition from oil to alternative sources of energy. The facts are plain. Current oil consumption continues to exceed new discoveries. The world has only a few decades in which to make this transition. Indeed, several countries of the Arab Gulf may have even less time in which to find an alternative source of national income. The Arab Gulf faces the vital challenge of transforming the flow of oil revenue over this short period into human and physical capital that can replace oil as a source of income, just as the industrialised world is trying to reduce its own dependence on oil.

There are undoubtedly very difficult years ahead. It is clear that we cannot afford a situation in which oil-consumers continue to waste energy while the oil-producing countries are unable to utilise a substantial part of their oil revenues.

The objectives of the two groups — both oil-producers and oil-con-sumers — are not necessarily divergent. On the contrary, their success in meeting the challenges facing them requires mutual trust and co-operation. Should they fail, the future might truly be characterised, as one Western oil expert has so profoundly put it, as the 'years that the locust hath eaten'.

Ladies and Gentlemen: Mutual trust emerges out of understanding circumstances and objective conditions, and out of a common apprecia-tion of the aspirations, vital interests and goals of all the parties involved. Our symposium has been a modest gesture in that direction.

The circumstances underlying recent oil price changes have, we feel, not been fully explained. OPEC oil-producing countries in general, and Arab oil-producers in particular, are frequently charged in some mysterious way with perpetrating shortages. We have, it is alleged, even cashed in on the industrialised countries' needs for energy. We have been accused of using oil as an instrument of political pressure. The list of charges even extends to include our responsibility for much of the economic stagnation recently endured by the Western world.

The truth of the matter is somewhat different. Most of these events are but symptoms of what is in actual fact a real world oil shortage. These events are also the outcome of the desire of the oil-producing nations to pursue policies which meet the needs of their people for a secure future. At the same time, the oil-producing nations wish to assert their own legitimate concerns as independent states about the current world economic and political order.

Ladies and Gentlemen: We in the Arab world, and in the Arab Gulf states in particular, pay considerable attention to the way events in the Arab world are received in North America and the West and other parts of the world. We sometimes feel that there is a limited understanding of our aspirations and achievements. For many years now we have been exerting a sustained effort to provide our people with a dignified and prosperous way of life. We have moved to integrate our industrial activities from crude oil production into refining. We have now gone further, into the production of petro-chemicals. We have established a strong regional association among the Arab Gulf states with the aim of fostering joint economic and social development, especially within the Gulf Co-operation Council region and the entire Arab world. We have erected an adequate social and economic infrastructure and a system of social benefits that is the admiration of even the most ardent reformers.

However, to be successful in our effort to extend the life span of our hydrocarbons beyond the gestation period of economic development we need a definite and substantial co-operative effort from the West, in general, and from Canada in particular.

We look to you as friends. And indeed there are many similarities between us. For in Canada you have enjoyed your own great experiment in federation. Each province has its own independent governing body. You have long subdued ancient quarrels and disputes.

We in the Arab Gulf, in fact, sought to benefit from the Canadian example. The Arab Gulf states consist of Saudi Arabia, Iraq, Bahrain, Qatar, the United Arab Emirates, Oman and Kuwait. We have, if you like, seven members of our closely-knit brotherhood and neighbourhood of nations. You in Canada have 12 provinces or territories from mighty Ontario to the relatively small Prince Edward Island.

The similarities are evident. We are seeking to reduce customs barriers, to make business together on equal terms. In the few short years since our concepts of Gulf co-operation began, we have made great strides for the welfare of all our people, and at every level of life.

Ladies and Gentlemen: We look to you for co-operation. Canada has ample natural resources and sophisticated technological skills and equipment. Canada has a world-renowned technology in the fields of nuclear reactors, communications and transportation. These are precisely what we in the Arab Gulf desire to acquire for our own future development.

For many reasons we feel very comfortable with Canada. After all, you do not carry the old stigma of imperialism. You are a pioneering and developing society. You are also a society concerned with human rights and fair play in international affairs.

The recent Canadian position on the rights of the Palestinians to a home-land and their inalienable right to self-destiny have, I assure you, been widely appreciated by the whole Arab people.

We would like you to support the peace initiative which aims at putting an end to the current Gulf war which has been raging now for four long years. Our brothers in Iraq have tried valiantly to end it. They need our co-operation to stop the bloodshed and to bring lasting peace to the region. Together, indeed, we could work for a lasting peace in the Middle East on the firm foundations of justice and liberty. The challenges are many but the rewards of Canada's co-operation with the Arab Gulf could be high and mutually beneficial.

The concept of mutual dependence inevitably transcends any economic dimension. It ultimately rests on the establishment of relations between people as members of the human family. We never forget in the Arab Gulf that we ourselves are part of the Third World. We try our best to extend a helping hand to our brothers in the rest of the Third World as much as we are able, but we also look to you to share this responsibility.

Ladies and Gentlemen: To conclude, we have certain objectives.

— To enlist your co-operation in the field of energy.
— To encourage mutual trust and understanding.
— To be partners in every aspect of progress.
— To work for peace with justice in Palestine and the Gulf.

This symposium has, we believe, contributed towards the achievement of these objectives.

Thank you, Mr President.

INDEX